Contemporary Activities in

ASTRONOMY

Contemporary Activities in

ASTRONOMY

A PROCESS APPROACH

FOURTH EDITION

Darrel B. Hoff
Jeffrey A. Wilkerson
Luther College

Kendall Hunt
publishing company

Book Team
Chairman and Chief Executive Officer Mark C. Falb
President and Chief Operating Officer Chad M. Chandlee
Vice President, Higher Education David L. Tart
Director of National Book Program Paul B. Carty
Editorial Manager Georgia Botsford
Senior Editor Lynne Rogers
Vice President, Operations Timothy J. Beitzel
Assistant Vice President, Production Services Christine E. O'Brien
Senior Production Editor Abby Davis
Permissions Editor Renae Horstman
Cover Designer Janell Edwards

Some labs in this work were acquired from *Activities in Astronomy* by Hoff, Kelsey, Neff.

Front Cover: This image shows the star and gas-rich region of the sky near the bright star Antares. Globular star cluster NGC 6144 just below and left of the center. Part of the larger, nearer globular cluster M4 peeks out above the blue bar across the bottom. These clusters are groups of old stars (see Exercise 28) that orbit the center of the Galaxy in a halo (see Exercise 29). Sigma Sco is the bright star appearing on the right edge of the field of view. It is both a binary star and a pulsating variable star. The Sh 2-9 nebula can be seen around it.

Kendall Hunt
publishing company

www.kendallhunt.com
Send all inquiries to:
4050 Westmark Drive
Dubuque, IA 52004-1840

Student Edition: ISBN 978-0-7575-6691-2
Special Edition: ISBN 978-0-7575-6699-8

Printed in the United States of America
10 9 8 7 6 5 4

CONTENTS

PREFACE

We are delighted to offer our popular astronomy laboratory manual in its updated fourth edition. In this and in previous incarnations, it represents over 30 years of service to the educational community.

New adopters of the book will notice that our book stresses as a meta-message the **processes** of science. We are pleased to see how its philosophy correlates to recommendations published by the Education Office of the American Astronomical Society under the leadership of Dr. Bruce Partridge, Haverford College.* Two national meetings were held in May and June of 2001 and leaders from roughly 30 research institutions drew up a list of goals for courses at the college level offered primarily for the non-specialist in astronomy.

Under the category of "Goals" (skills, values, and attitudes) we find the following:

1. Students should be exposed to—
- The excitement of actually doing science
- The evolution of scientific ideas (science as a cultural process)

2. Students should be introduced to how science progresses, and receive training in—
(Italics added to indicate themes present in many of the activities in this manual.)
- *the roles of observation, experiments, theory and models*
- *analyzing evidence and hypotheses*
- *critical thinking (including appropriate skepticism)*
- *hypothesis testing (experimental design and following the implications of a model)*
- *quantitative reasoning (and the ability to make reasonable estimates)*
- *the role of uncertainty and error in science)*
- *how to make and use spatial/geometrical models*

3. And we should leave students—
- more confident of their own critical faculties
- inspired about science in general and astronomy in particular
- better equipped to follow scientific arguments in the media
- For details about this study see Commission 46—Astronomy Education and Development of the International Astronomical Union Newsletter 57, October 2002.

ACKNOWLEDGEMENTS

Thanks are due to all of those many people who contributed in various ways to previous works. First and foremost, thanks to Dr. Linda Kelsey, Dr. John Neff, and the late Dr. James Van Allen for their early work at the University of Iowa. Thanks also to those dozens of students who provided assistance and insight for this manual.

Special thanks go to Dr. Bruce Palmquist for creating the Instructor's Manual for this book.

ABOUT THE AUTHORS

Darrel Hoff was born on a small farm in western Wisconsin and credits his mother with his early interest in astronomy. Growing up, during the pre-electricity days in rural Wisconsin, his mother would lie on the lawn with the family on summer evenings and tell stories about the stars and constellations. His early education took place during the Great Depression in a rural one-room school. He later attended high school in Viroqua, Wisconsin and college at Luther College in Decorah, Iowa. Subsequently he taught high school chemistry and physics, served two tours of duty with the U.S. Army, and completed a master's degree at the University of Northern Iowa and a Ph.D. at the University of Iowa.

Darrel was a professor of Astronomy and Science Education at the University of Northern Iowa for many years then moved East to conduct a number of teacher-enhancement programs at the Smithsonian Center for Astrophysics at Harvard University. His research interest is in variable stars and he has numerous publications in that field. He is widely published in science education as well.

After leaving Harvard, Darrel and his wife, Ardy, retuned to Iowa to retire nearer family. Darrel was re-employed as an Adjunct Professor of Physics and Science Education at his alma mater, Luther College. He and Ardy, who is an art educator have two children and two grandchildren. In addition to astronomy, Darrel enjoys gardening, carpentry work, and reading about military history.

Jeffrey Wilkerson developed an interest in astronomy by watching the dark skies while growing up in southern Indiana. He spent as much time as possible outdoors and was drawn naturally to observe and study the night sky. Jeff's interest in astronomy was nurtured by Bob Kasting, a caring high school physics teacher who started his day early to allow students to work in the school planetarium before classes. In addition, he made the physics laboratory and equipment available to students wanting to work outside class. One piece of equipment in particular, a homemade 6-inch reflecting telescope, fueled Jeff's interest in astronomy.

Jeff attended Indiana University, where he studied physics and worked in research groups studying cosmic rays and the properties of atomic nuclei. He earned his Ph.D. from the Berkeley campus of the University of California where he developed instrumentation for X-ray astronomy. In 1997, he joined the faculty of Luther College where he teaches physics, astronomy and general science. He has done collaborative research work with more than thirty undergraduate students in the past ten years. These projects have included monitoring variable stars in the field of M23, searching for period evolution in eclipsing binary stars, and monitoring the long-term luminosity stability of a wide variety of stars.

INTRODUCTION

The Processes of Science

The common view of science is that it consists of a body of facts; a compendium of ideas and a set of dogmatic concepts. Typically, in beginning science courses we reinforce this idea by the manner in which we tech the material and how we test over science as a product.

Science as a Product

For example, we catalog the number of planets and moons in the solar system, we describe the surface of Mars, and we list the features seen on the surface of the sun. Then we ask students to regurgitate this information. In so doing, we communicate that science is a product. In reality, science is only partly product.

Science as a Process

As practiced by most scientists, science is a *process*. This word describes science as a vital, ongoing enterprise. The word conveys the impression that science is dynamic and not static. It says that science is a "doing profession" and that it is in constant change. Unfortunately, this aspect (perhaps the most crucial aspect) is seldom communicated to the beginning student.

What do we really mean by "process" as part of science? It usually refers to those intellectual skills and practices that when taken as a group, characterize science. These skills may not be unique only to science, but in aggregate they are more characteristic of science than any other intellectual discipline.

Table I.1. Fifteen Science Processes identified by The Center for Unified Science Education at Ohio State University

• Classifying	• Hypothesizing
• Inventing Concepts	• Using Numbers
• Designing Experiments	• Controlling Variables
• Questioning	• Observing
• Identifying Variables	• Formulating Models
• Interpreting Data	• Using Logic
• Defining Operationally	• Inferring
• Predicting	

How to Use This Laboratory Manual

There are many exercises contained in this lab manual—more than can possibly be completed in a single course. This abundance of exercises allows each instructor to build a laboratory curriculum to fit the specific needs of their course.

In this book, attention will be regularly drawn to the processes of science identified by the Center for Unified Science Education. (See Table I.1) it will be done in three ways.

1. On occasion instructors will comment on how a certain idea was arrived at, as well as discuss the idea itself.
2. Students practice these skills by completing experiments.
3. Students should watch for examples of the "process" aspect of science while reading the text and course lecture materials.

Features of This Manual

Each exercise follows the same basic format—Purpose and Processes, References, Introduction, Procedure, Exercise Worksheets, Discussion Questions, and Personal Vignettes.

- Purpose and Processes—this feature indicates what knowledge will be gained from completing the exercise and states which science processes are featured in the exercises.
- References—references are included at the beginning of the exercise as necessary to understand the material.
- Introduction—the introduction provides background information needed to complete the exercise.
- Procedure—the procedure details the step-by-step process for completing the assignment.
- Exercise Worksheets—each exercise contains an assignment for students to complete as they work through the exercises.
- Discussion Questions—at the end of some exercises, discussion questions provide students with a collaborative learning environment. These questions can also be given as homework assignments.
- Personal Vignettes—notable individuals who have contributed to the scientific process of astronomy are highlighted in these vignettes.

Web Links

The standard paper and pencil labs are retained in this fourth edition for courses where equipment and computer access are limited.

Web links are provided to permit many of the labs to be done with information downloaded from the Web.

The lab manual is intended as a stand-alone resource for use in a laboratory setting.

Those who are interested in further information can find extensions to some activities on the authors' website available at http://faculty.luther.edu/~wilkerje/astronomy.html

Included are links to images and additional data for use in some exercises. There are also links of general interest and a handful of web-based activities to complete. A password is not needed to access the authors' website.

Color Plates

There are 16 full color plates in this book to give students accurate color descriptions of processes they are studying.

Appendices

The appendix contains valuable information to help students gain knowledge and insight to astronomy.

Appendix 1: How to Write Laboratory Reports gives students step-by-step instructions for preparing lab reports.
Appendix 2: How to Handle Data helps the beginning astronomy student understand the importance of graphing and how uncertainties play a role in experiments.
Appendix 3: Equipment Notes provides detailed information about equipment needed to complete each exercise.
Appendix 4: SC1-SC2 Charts, Star Finder, and Translucent Graph Paper

- SC001 and SC002 Constellation Charts are included for referencing information.
- Star Finder in included in this book. It adjusts for time of day and time of year, helping students locate objects.
- Graph Paper is included for use in working exercises.

PART I
OBSERVING THE SKY

Visual Astronomy

Purpose and Processes

The purpose of this exercise is to become acquainted with the night sky, to locate some of the more prominent stars and constellations, and to become familiar with some of the literature that is available to assist in naked-eye observing. A series of sample star counts will be made in order to estimate the total number of stars on the celestial sphere that are visible to the naked eye. The processes stressed in this exercise include:

Observing
Using Numbers

References

In addition to the star charts and star finder included in this manual and in your text, there are several other sources of information that are useful in visual observation. These references may be divided into six categories: star finders, periodicals, field books, annuals, atlases, and CD ROMs. Several are discussed in greater detail in Exercise 3 *Star Charts and Catalogues.*

STAR FINDERS

Since stars exhibit an annual motion (due to the earth's revolution about the sun) and a diurnal motion (due to the earth's rotation on its axis) the locations of stars change from day to day and hour to hour, making the orientation of star charts difficult. Star finders can be adjusted for both the time of day and time of year, making the task of locating objects easier. (See Appendix 3 *Equipment Notes.*)

PERIODICALS

The positions relative to the stars of celestial objects such as planets and asteroids change from year to year so that they can not be easily located without consulting a current source of information. Four periodicals that list planetary positions and current celestial events, provide up-to-date information about other subjects of interest to astronomers, and that are commonly used by the beginning student are:

- *Astronomy* [Monthly]
  AstroMedia Corporation
  21027 Crossroads Circle
  Box 1612
  Waukesha, WI 53187-1612
- *Publications of the Astronomical Society of the Pacific* [Monthly]
  Astronomical Research Journal
  University of Chicago Press
  1427 E. 60th St.
  Chicago, IL 60637-2954
- *The Planetary Report* [Bimonthly]
  The Planetary Society
  65 North Catalina Ave.
  Pasadena, CA 91106
- *Sky and Telescope* [Monthly]
  Sky Publishing Corporation
  90 Sherman St.
  Cambridge, MA 02140-3264

FIELD BOOKS

These publications provide more comprehensive information about what can be seen in the sky with and without optical aid. They generally provide information about the history of the constellations and the objects in them, nebulae, double stars, colors of stars, and variable stars. Recommended resources are:

- Pasachoff, J., et al. (1999). *Field guide to the stars and planets.* Boston: Houghton-Mifflin, Harcourt.
- Whitney, C. (2006), 5th Ed. *Whitney's star finder.* New York: McGraw-Hill Companies.

CD-ROMS

Sky watching has come to depend heavily on electronic aids—computer programs and CD-ROMs. Programs are available to calculate the appearance of the sky for any given latitude and longitude for any given date and time. Most can be directly linked to modern small aperture telescopes for accurate location and guiding. Some list the information requested or give a graphic display of the sky. Some will take you to other planets in the solar system to illustrate the motions of its moons. These capabilities are dependant upon the size of the RAM and the speed of the microprocessor in the computer. Only a few are listed below. Others are advertised in *Sky and Telescope* or *Astronomy* magazines. The Astronomical Society of the Pacific and Sky Publishing Corporation catalogs list the most recent versions and current prices.

See the website

www.seds.org/billa/astrosoftware.html

for a more complete listing of such programs. At the publication time of this book, about 100 programs were listed including many available for free. A late and valuable addition to this field is the availability of programs for palm pilots.

- *Voyager III 3.0* CD-ROM for Macintosh or Windows
  This is still the most popular interactive planetarium program. It is unequaled in speed, accuracy, versatility, and ease of use. Contains 259,000 SAO star, and 20,000 deep-sky objects.

- *The Sky* Version 5 CD-ROM for Macintosh or Windows
  Features include the Hubble Guide Star Catalog's 19 million stars and non-stellar sources, and 750 color images of the planets and deep sky objects.
- *Realsky* for Windows
  Now you can have the most accurate digitized sky survey right on your desktop. Stars as faint as the 19th magnitude are shown with an angular resolution of less than 2 arc seconds.
- *Starry Night Pro* CD-ROM for Macintosh or Windows
  Extremely realistic and visually stunning program. Serious observers will enjoy the Hipparcos/Tycho star databases and other advanced features.

ANNUALS

Another useful type of publication is an annual or calendar, which is published yearly and provides some of the information found in periodicals and field books. They give positions of the planets, phases of the moon, rising and setting times of the sun and moon, locations and properties of the stars, etc. Some commonly used annuals are:

- Bishop, R. (Ed.). *The observer's handbook.* Toronto: Royal Canadian Astronomical Society.
- Ottewell, G. *Astronomical Calendar.* Greenville, SC: Furman University.

Various wall calendars showing historical and celestial events are available from AstroMedia, Sky Publishing, and Optica. Information about celestial events for an upcoming year can also be found in different "farmer" almanacs available at local drugstores and bookstores in October or November of the previous year. Addresses are provided in Appendix 3.

ATLASES

A fifth type of publication is an astronomical atlas. Atlases have traditionally been published in soft-cover versions and still are, but now some are available on diskettes.

- Tirion, W. *The Cambridge star atlas* (3rd ed.). Cambridge, UK: Cambridge U. Press.
  This large-format hardcover atlas gives pairs of months with all-sky maps which orient observers to the visible sky at different times of the year.
- H.J.P. Arnold, Paul Dobherty, & Patrick Moore. *The photographic atlas of the stars.* Institute of Physics.
  This hardcover atlas presents 45 pairs of wide-field astrophotos taken from both the Northern and Southern Hemispheres. Each is faced with a detailed map made from negatives of the corresponding photo. The negative maps are labeled with the best-known stars and deep-sky objects.
- Ridpath, I. (Ed.). (2000). *Norton's 2000.0 star atlas and reference book* (20th ed.). Benjamin Cummings Publisher.
  This perennial favorite is regarded by many as the only essential reference guide needed by a beginning observer of the sky. Its value is testified to by the fact that it has gone through so many editions.
- Hirschfeld, Sinnott & Ochsenbein. *Sky catalogue 2000.0.*
  Vol. 1: *Stars to magnitude 8.0* (2nd ed.). Cambridge, UK: Cambridge U. Press.
  Vol 2: *Galaxies, Double and variable stars and star clusters.*

Introduction

1. CONSTELLATIONS

The Greeks recognized the impossibility of attempting to learn much about the heavens without first organizing their information about the vast number of stars in some systematic manner. The geometric arrangement of some stars provided the Greeks with a natural organizational system that we call the *constellations.* These are accidental groupings of stars whose outlines have been given the names of people, objects, and animals they were thought to resemble or chosen to represent. Although the modern astronomer no longer employs these constellations as the ancients did for mythological or astrological purposes, he does use them for quick reference purposes. A number of bright stars are named in the Bayer System making use of the constellation names, and current celestial events are frequently given names of the constellation in which they occur. Novae, for example, are named for the constellation in which they are observed and the year of their occurrence (such as Nova Serpens-1970 or Nova Puppis-1919).

Originally the constellations did not encompass all of the stars, but in 1928 the International Astronomical Union divided the entire celestial sphere into 88 constellations using regular north-south or east-west boundaries so that all stars (and areas of the sky) are now assigned to a constellation.

2. STAR COUNTS

Further observations can be made with the naked eye to study the structure of the universe. We can try to determine the total number of stars visible to the naked eye and visible in telescopes of various sizes, the distribution of stars on the celestial sphere, and the relative number of stars of various magnitudes.

To actually count the stars we can see would be an impossible task. We can simplify the problem by counting the number of stars in several small regions of the sky. To estimate the total number of stars visible, we can calculate the ratio of the area of these regions to the area of the whole celestial sphere. A cardboard tube can be used for observation, and the area of sky being observed can be calculated from the length and radius of the tube. (See Exercise 2 for procedure.)

Procedure

Pull out the two cardboard sheets in Appendix 4. Cut out, fold, and paste **PART 1** of the star wheel holder. Cut out the star-date wheel from **PART 2** and slip it into the holder. The shell should rotate freely, matching dates with times on the holder. By matching the time of night with a specific date, the stars visible in the sky for that date and time are visible in the oval window. This view is for a mid-northern latitude. The center of the opening represents straight-up or zenith. The edge of the opening represents the horizon with the cardinal directions indicated (Figure 1-1).

Example:

Rotate the star wheel until midnight is pointing at December 10. The constellation Orion, with the three stars in his belt, is about halfway between the southern horizon and zenith; the brightest star in Auriga is close to zenith (use the SC2 chart in Appendix 4 to identify this star as Capella); and Ursa Major is low in the north-northeast with the handle of the Big Dipper pointing toward the horizon.

As you rotate the star wheel (counter-clockwise), notice that stars are rising from the eastern horizon and setting toward the western horizon. You should also notice that there is a

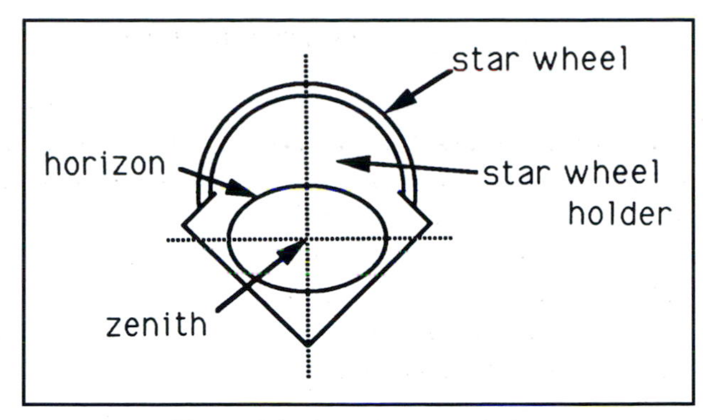

Figure 1-1. Completed star finder.

group of constellations in the northern sky that are not rising or setting, they move in circles around the north celestial pole (close to Polaris). These stars that never rise or set are called "circumpolar stars." To study circumpolar stars more comfortably, hold the star finder with *NORTH* down.

You should be able to locate the common circumpolar constellations and the constellations that are associated with each season of the year. Since this exercise is often done in the fall, some general techniques will be discussed to assist in locating the circumpolar constellations and to locate the constellations of fall. Similar techniques can be used for other seasons. Table 1-1 lists the constellations and bright stars you should be able to locate each season. Use this star finder in the following problems, matching the date with the time given at the beginning of each section. Use the SC1 and SC2 charts in Appendix 4 to identify specific stars in a given constellation.

Table 1-1. Constellations and Objects in Them

Circumpolar Constellations	Auriga (Capella)
Ursa Major (Alcor, Mizar, Duhbe, Merak)	Taurus (Aldebaran, Pleiades Cluster)
Ursa Minor (Polaris)	Orion (Betelgeuse, Bellatrix, Rigel)
Cassiopeia (Caph)	Spring Constellations
Cepheus	Gemini (Pollux, Castor)
Draco (Thuban)	Canis Minor (Procyon)
Fall Constellations	Canis Major (Sirius)
Lyra (Vega)	Cancer (Beehive Cluster)
Cygnus (Deneb)	Leo (Denebola, Regulus)
Aquila (Altair)	Hydra
Delphinus	Corvus
Sagitta	Summer Constellations
Pegasus	Coma Berenices
Capricornus	Boötis (Arcturus)
Aquarius	Libra
Winter Constellations	Scorpius (Antares)
Andromeda (M31)	Sagittarius
Aries	Corona Borealis
Pisces	Serpens
Cetus	Ophiuchus
Perseus (h and χ clusters)	Hercules (M 13 Cluster)

CIRCUMPOLAR CONSTELLATIONS FOR A NORTHERN MID-LATITUDE OBSERVER

It is easier to locate a celestial object if some prominent constellation can be found and used as a reference point. Ursa Major (the Big Dipper) serves as a good reference point. Rotate the star wheel so that 9:30 P.M. points at September 15th. Ursa Major lies nearly parallel to and quite close to the northern horizon. The stars at the end of the dipper's cup will be to the east and the cup will open upward. The two stars farthest to the east are called the "pointer stars." They are separated by five degrees of arc (the approximate width of a closed fist held at arm's length). Use these two stars and trace a line upward for a distance of about 30° (or six "fists") to locate the brightest star in Ursa Minor (the Little Dipper). Ursa Minor lies nearly parallel to Ursa Major, and the dippers open toward each other. The brightest star in Ursa Minor is called Polaris and its location is near the north celestial pole.

Locate the pointer stars in Ursa Major, follow them across Polaris to about 30° on the other side of the pole from Ursa Major, and find a large "W" lying on its side. This is the constellation Cassiopeia. Winding its way between the two dippers is an irregular line of stars bearing the name of Draco (the dragon). The fifth circumpolar constellation is composed of an irregular square topped by a triangle of stars, and is called Cepheus. In the autumn Cepheus is located above the pole and west of Cassiopeia with its point directed toward Polaris.

Locate all the stars and constellations on the circumpolar list (Table 1-1) and answer the following questions.

1. Examine Zeta Ursa Majoris (the next-to-the-end star on the handle of the dipper) carefully with the naked eye, and if possible, with binoculars. What do you see?
2. Compare the brightness of the stars in Ursa Major. What observation can you make about them?
3. Compare the brightness of α Ursa Minoris with the second star in the handle of the "dipper." What observations can you make?
4. If Ursa Major is lying parallel to the horizon and below the pole at 10:00 P.M. on September 14th, what will be its orientation at 4:00 A.M. on September 15th, and why?
5. Locate β Cassiopeia. This star has the proper name Caph. A line drawn through the pole star, passing through Caph, and continued southward intersects the Vernal Equinox. What is the significance of this part of the sky with respect to our seasons? (Refer to a star chart if necessary.)

CONSTELLATIONS OF FALL

Turn to face the southern horizon and the region of the sky overhead. At about 9:30 P.M. (standard time) in mid-September there are three bright stars that form the apexes of a large right triangle. The brightest of the three is Vega (in the constellation Lyra), which is farthest west and is at the right angle of the triangle. The other two stars are Deneb (in Cygnus, the Northern Cross) and Altair (in Aquila, the Eagle). These three constellations make good reference points for the rest of the fall sky. Note in the eastern sky a large area defined roughly by four stars arranged in a great square. This is the body of the constellation Pegasus, which is connected to Andromeda.

1. Locate the stars Vega and Altair. Compare and describe their colors. What do you think their colors indicate?
2. Locate Vega and near it the star Epsilon Lyrae. Examine this star with binoculars if possible. What do you observe?
3. If a small telescope is available, locate Beta Cygni (called Albireo). Examine the star with the telescope and report your observations.
4. Are any planets visible? If so, identify them and locate their positions with respect to the brighter stars. Refer to a star chart to estimate the right ascension and declination of each planet. Compare the planets' brightnesses to that of the brighter stars.

5. Is the moon visible? If so, what is its phase? Locate its position with respect to the brighter stars using a star chart from Appendix 4. Estimate its right ascension and declination.

CONSTELLATIONS OF WINTER

Face south or southeast at about 9:30 P.M. in January and look for a group of three stars of almost equal brightness in a line. There is a second dimmer line of three stars below and at an angle to the first line. This marks the belt and sword of Orion. This constellation makes the best reference object for the winter sky.

The belt points up and to the right toward the bright star Aldebaran in Taurus. It points down and to the left toward the bright star Sirius in Canis Major. Taurus resembles a large "V" in shape, one side of which points up to the constellation Auriga and the other side of which points to Gemini. Locate all the winter constellations listed in Table 1-1 and answer the following questions.

1. Make a small table and enter descriptions of the colors of Sirius, Rigel, Betelgeuse, Capella and Aldebaran.
2. Locate Sirius, Betelgeuse and Procyon. Describe the configuration they make and carefully compare their colors.
3. Carefully compare the brightness of Sirius and all the other bright stars of the winter sky. What do you conclude?
4. Use binoculars to examine the middle star in the sword of Orion. What do you see?
5. Use binoculars to examine the Pleiades. Describe your observations.
6. Is the moon visible? If so, what is its phase? Locate its position with respect to the brighter stars and estimate its right ascension and declination.

CONSTELLATIONS OF SPRING

In late spring look west at about 9:30 P.M. and locate two bright stars approximately parallel to the western horizon. These are Castor and Pollux, the two brightest stars in Gemini. Now turn south and look for a backward question mark. This marks the forequarters of Leo the lion, with Regulus as its brightest star. If you have looked at the winter sky, notice how comparatively empty the sky seems in spring.

Locate the spring stars and constellations listed in Table 1-1 and answer the following questions.

1. Locate Regulus and Denebola in Leo. Compare their colors and relative brightnesses.
2. Locate the constellation Cancer. Examine that region of the sky with binoculars. What do you see?
3. Locate the triangle that makes up the hind quarters of Leo. Draw a perpendicular line from the hypotenuse of the triangle outward from Denebola and examine the region with binoculars. What do you see?
4. If a small telescope is available, examine α Canis Venaticorum and α Leonis. Compare the two observations.
5. Is the moon visible? If so, what is its phase? Locate its position with respect to the brighter stars and estimate its right ascension and declination.

CONSTELLATIONS OF SUMMER

Locate Ursa Major and trace a line outward from the handle of the "dipper" southward until it intersects a bright yellow star. This is Arcturus in Boötis. Continue the curved line southward until it intersects a second bright (blue) star named Spica in Virgo. Now look low along the southern horizon for a fishhook-shaped figure, Scorpius. Immediately to the east of Scorpius is Sagittarius, marked by a small inverted dipper-shaped group of stars. To the east of Sagittarius is a "delta"-shaped grouping of stars outlining Capricornus. The four constellations

Virgo, Scorpius, Sagittarius, and Capricornus mark one-third of the zodiac. A fifth, Libra, can be located with some difficulty between Virgo and Scorpius.

1. Compare the colors of Spica and Antares.
2. If you have binoculars, slowly scan the area above the "hook" of Scorpius. What do you see?
3. Locate Corona Borealis. Compare the relative brightnesses of the stars in this constellation.
4. If the sky is dark, carefully scan the Milky Way with binoculars and record your observations.
5. Is the moon visible? If so, what is its phase? Locate its position with respect to the brighter stars and estimate its right ascension and declination.

Discussion Questions

1. What was the general condition of the sky while you were making observations? Was the moon visible? How might these factors affect your results? Are there any other local factors that should be considered?

2. How and why might the altitudes of the areas of sky used for counting stars affect your results?

Exercise 2

Observing Exercises

Purpose and Processes

The purpose of this exercise is to observe many different solar system and deep-sky objects. The processes stressed in this exercise include:

Using Numbers
Observing
Using Logic
Inferring
Interpreting Data

References

1. *Norton's Star Atlas and Reference Book,* Ridpath, Benjamin Cummings, 2003.
2. *Sky Atlas 2000.0*
3. *Uranometria 2000.0 Vol. 1 and 2*
4. *A Field Guide to the Stars and the Planets,* Pasachoff, Peterson and Tirion, Houghton-Mifflin, 1999.
5. *The Observer's Handbook*
6. *Sky Gazer's Almanac*
7. *Astronomical Calendar*
8. *The Astronomical Almanac* (current year)
9. *Sky and Telescope* (current issue) or *Astronomy* (current issue)

Introduction

The sky offers a variety rewarding opportunities for observers with a variety of observing equipment. In general, solar system objects can be relatively easily observed. The positions of the brighter planets against the stars can be tracked with simple angle-measuring devices, as can the position of the moon. A telescope allows one to observe the fainter planets Uranus, and Neptune and opens craters on the lunar surface for study. Sunspots can be observed with use of a solar filter or via image projection. NEVER

LOOK AT THE SUN THROUGH A TELESCOPE unless adequate precautions are taken. The telescope allows many nebulae, galaxies, star clusters and double stars to be examined. The careful use of comparison stars allows observers to detect stellar variability. This exercise suggests many observing opportunities, with or without the aid of a telescope. Observation data sheets are provided. Books and magazines listed in the references and elsewhere are valuable resources in aiding the beginning sky observer. Positions of the planets can be found in magazines, The Astronomical Almanac and with the use of planetarium software (e.g., *The Sky*). These same resources provide information on the lunar phase, important both for knowing when the moon is available for study and when too much moonlight is likely to ruin the observation of more subtle objects like nebulae and galaxies. Double stars, star clusters, galaxies and nebulae can be located with the aid of books such as Norton's Star Atlas and Reference Book, Sky Atlas 2000.0, Uranometria 2000.0 and planetarium software.

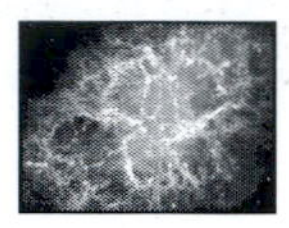

Procedure

THE MOON

1. Set up and align the telescope (see Appendix 2) and start the clock drive if one is available. Since the moon is moving with respect to the stars, the telescope will not drive at exactly the right rate. However, use of a clock drive will allow the moon to remain in the field of view for a longer time.
2. Record the date, time, location, and general weather and sky conditions for your observations.
3. Estimate the altitude and azimuth of the moon, and indicate its position relative to nearby bright stars.
4. Observe the moon with the naked eye and do a quick sketch showing as much detail as possible. Name its phase.
5. Observe the moon with a low power (long focal length) eyepiece. Compare its markings to your naked eye sketch. How does the orientation of objects seen through the telescope compare to that seen with the naked eye?
6. Record the focal length or magnification of the eyepiece and sketch the lunar surface or parts of it in as much detail as time permits. Look for and sketch as many different types of features as you can see.
7. Compare your sketch to a lunar map such as that found in *Norton's 2000.0,* a textbook, or Figures 2-1 and 2-2 and name the major maria and craters on your drawings.

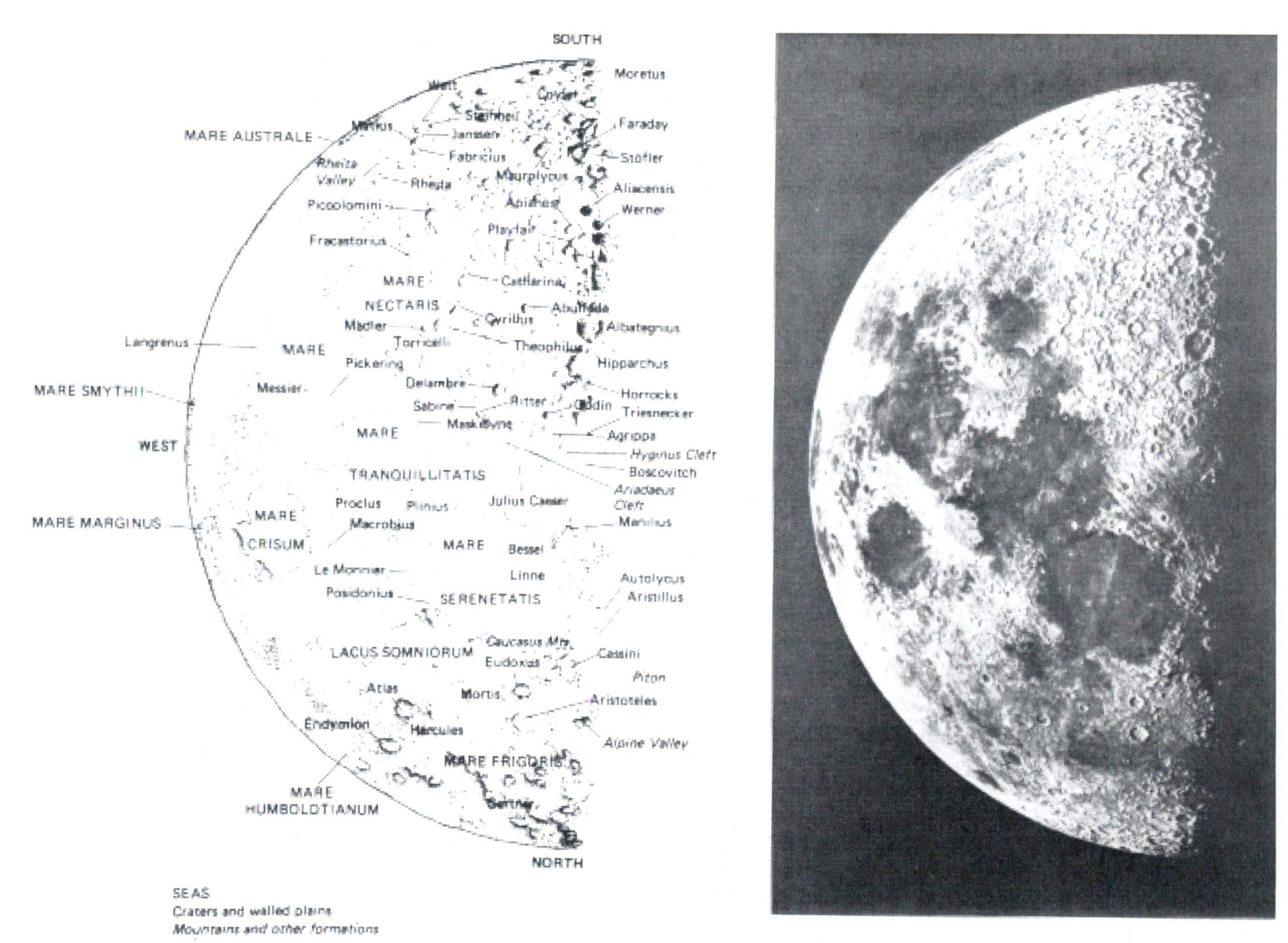

Figure 2-1. First-quarter moon (telescopic view). Image on left from *Exploration of the Universe,* Updated Brief Edition by George Abell, copyright © 1973 by Holt, Rinehart and Winston, reproduced by permission of the publisher. Image on right from The Observatories of the Carnegie Institution of Washington.

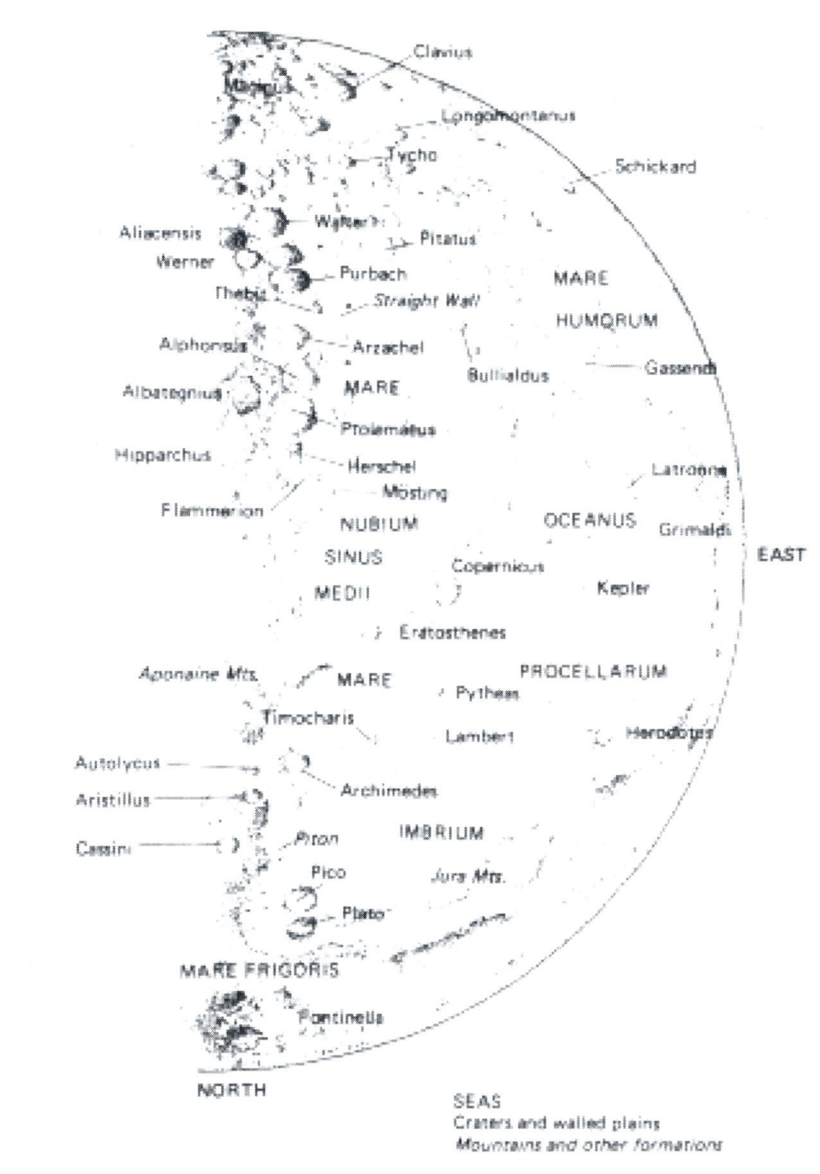

Figure 2-2. Third-quarter moon (telescopic view). Image on left from *Exploration of the Universe,* Updated Brief Edition by George Abell, copyright © 1973 by Holt, Rinehart and Winston, reproduced by permission of the publisher. Image on right from The Observatories of the Carnegie Institution of Washington.

Thoughts from Dan Durkin . . .

Observing the Sky Through His Own Creation

When I was a small boy in the Maryland countryside, I remember sitting on the front steps with my mom and she would point out the Big and Small Dippers. As I grew up, went to college, and then on to graduate school to study physics, it's somewhat embarrassing to admit that my astronomy knowledge did not progress much beyond also recognizing Orion. I was fortunate, though, to come across John Dobson, as have many people who walk the streets of San Francisco at night have. On those rare nights when the fog is held at bay and the planets and the brightest stars shine through the city glare, John and his fellow sidewalk astronomers set up on street corners with their cannon-like Dobsonian reflector telescopes, beckoning all that pass by: "Come look at YOUR moon!;" "Saturn! Saturn! Come see Saturn!;" "Moons of Jupiter!."

The time had come to correct a tremendous vacuum in my education, and so it was that I signed up for John's class at the San Francisco Academy of Sciences—"How to make a telescope." I will never forget that view on the night of my first class. Previously built telescopes were set up on the Academy's steps as the San Francisco fog creeped over the hills and through the trees, threatening to obscure the crescent moon that reigned above. It was a race to catch a glimpse! I cautiously approached the eyepiece, inquisitively studied the homemade cardboard tube focuser, knelt down in homage, and peered through. What a tremendous sight! There it was, glimmering before me, slowly cruising through the field of view. I could reach out and touch it! Some parts were smooth and dark, and I would later learn these were mares, areas where molten moon rock flowed forth on the moon's surface. Other areas were quite battered. There were craters within craters, evidence of a violent history. Most amazing to me were the peaks of light along the light-dark terminator, places where only high mountainous areas caught the sunlight to send down to me on earth.

That first night was in October. By December I had constructed my own Dobsonian: a 10-inch mirror mounted in a 6-foot long tube that barely fit in my Saturn sedan. It's ironic that the moon was my first astronomical object to view, and still my favorite. A telescope that big beckons for the dark sky and the faint fuzzies that the bright moonlight overwhelms. It wasn't until February when the weather cleared, and a window opened to head towards the Sierras for the first real test. Tutored by two experienced amateur astronomer friends, we packed up the cars bound for Blue Canyon, a small air strip in the Sierra mountains that also played host to astronomers. That night, though, there was only one other telescope there, in the observatory built by a local group. The ground was covered by over three feet of snow, thankfully covered with a hard crust. The road was not plowed all the way in, so we had to park several hundred yards away and walk our behemoths in, occasionally poking through the hard crust and slipping through the other three feet to the frozen ground below. It was going to be a well-below freezing night, but I was warmed by the excitement of seeing wondrous new objects.

The telescopes were set up on the frozen tarmac, pointed skyward. As the sun set, the temperature plummeted, but something else miraculous happened. A bright light here, another there. The stars slowly revealed themselves until the sky was filled with them. A faint glow stretching across the southern hemisphere was also visible, reflections from dust in our solar system. This was a whole new world for me to explore!

I saw many objects that evening through my new telescope. Galaxies, exploding stars, remnants of exploding stars, Jupiter, and clusters of stars. Even meteors streaked over head. All firsts and all spectacular. One object stands out, though. In Canis Venatici, close to the Big Dipper's handle there is an object called M51, the Whirlpool Galaxy. Not one, but two galaxies. And one seems to be whirling into the other in galactic cannibalism. It's truly an amazing sight and it has never looked better than it did on that first night. I and my two friends were transfixed by it late that evening, each by our own telescope on the frozen tarmac. I had at least temporarily forgotten my frozen toes, it was such an unbelievable view. The silence was broken though in dramatic fashion as all three of us simultaneously exclaimed "wow!" at the red and green airplane lights that flew right through the Whirlpool.

Dan Durkin
Physics Graduate Student
Amateur Astronomer

THE SUN

Projection Method

1. Before pointing the telescope at the sun it is necessary to stop down the aperture of the telescope by taping a cardboard ring over the end of the tube as shown in Figure 2-3. UNLESS THIS RING IS USED TO BLOCK SOME OF THE LIGHT YOU MAY DISTORT OR MELT THE EYEPIECE as the telescope collects the sunlight and focuses it through the eyepiece! Cover the end of the finder telescope with an opaque cover to avoid the possibility of looking at the sun through it.
2. Point the telescope toward the sun by aligning the telescope along its shadow. Hold a screen (a piece of paper or cardboard) several inches from and perpendicular to the axis of the eyepiece (Figure 2-4). Center the image on the screen and focus the sun as clearly as possible.
3. Hint: In order to avoid oblate solar images the screen must be placed perpendicular to the axis of the eyepiece. If photographing the image, the camera should be held with the camera back approximately parallel to the screen. It may be helpful to center the sun's image in a square drawn on the screen, and align the image in the circular pattern of the field of view of the camera.
4. Use a telescope drive if one is available. Although you won't be able to align it exactly on the pole, an approximate alignment will allow the sun to remain in the field for a longer period of time.

Solar Filter

NEVER LOOK AT THE SUN THROUGH A TELESCOPE UNLESS ADEQUATE PRECAUTIONS ARE TAKEN.

1. There are two general types of filter: those that cover the end of the telescope tube, and those contained in an eyepiece. Check with your instructor for the characteristics and specific directions for the filter you will be using.

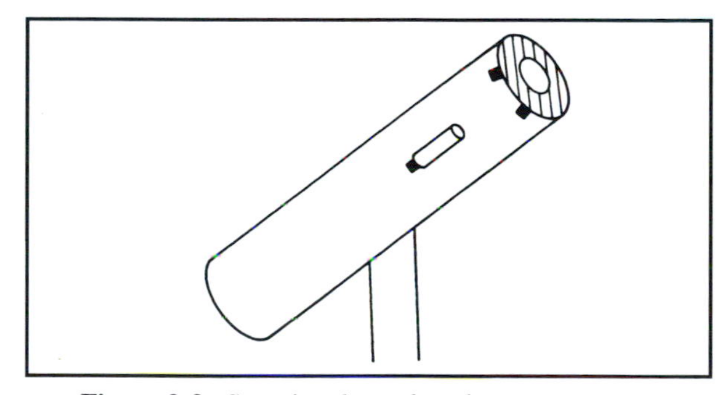

Figure 2-3. Stopping down the telescope aperture.

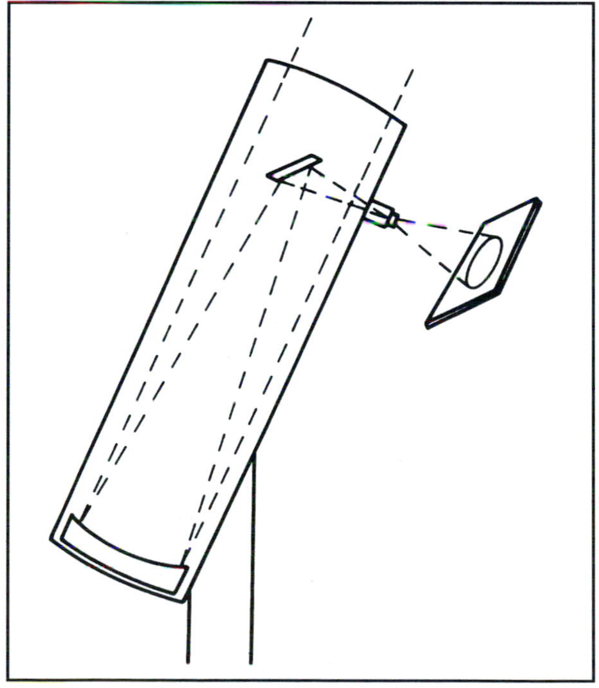

Figure 2-4. Sunspot projection.

2. Point the telescope toward the sun by aligning it along its shadow.
3. If your equipment adapts for photography, attach the camera and focus it as accurately as possible.
4. Use a telescope drive if one is available. Although you won't be able to align it exactly on the pole, an approximate alignment will allow the sun to remain in the field for a longer period of time.

Observing the Sun

1. Note the time, date and place of your observations as well as the general sky conditions (hazy along the horizon, some scattered thin clouds, etc.).
2. Photograph the sun or sketch its disc in as much detail as possible. Show the number and position of the sunspots. Sketch any other features that may be visible and tell what they are.
3. Observe the sunspots on as high a power as possible and sketch in detail their appearance. Indicate which spots you are observing on your observing sheet drawing of the low magnification image of the whole sun.
4. Look at an image of the sun as closely as possible and note its color or shading. Is the color uniform across the disc? Do you notice any bright areas? If so, sketch them. (A filter works best for this activity.)
5. If possible, come back a few days later and re-observe the sun, or compare your notes with someone who did the lab several days before or after you did. (Tell whose data you are comparing with.)
6. Print the best photograph for each day of solar photography. Identify the sunspots or groups on each picture or sketch, and determine the axis of solar rotation. Measure the solar latitude of each group of spots.
7. If the solar image is oblate on your prints, devise a mathematical method to correct for this in locating each sunspot group's direction and amount of motion for each time interval observed. Determine the period of solar rotation for each major sunspot or sunspot group.

TELESCOPIC OBSERVATIONS

1. Before coming to lab, use any of the references below to find the positions and record the coordinates of the planets.
2. Set up and align the telescope (see Appendix 2) and start the clock drive if one is available.
3. Record the time, date, location, and general weather and sky conditions for your observations.
4. Estimate the altitude and azimuth, as well as the time at the beginning and end of each planet's observations. Sketch any nearby stars and locate the planet on your sketch.
5. View the planet with the telescope starting with low magnification (a long focal length eyepiece) and then going to higher powers by selecting eyepieces with shorter focal lengths. Record the magnification or focal length of the eyepiece being used for each sketch or observation.
6. Sketch and describe what you see in as much detail as possible. Note any visible surface markings, colorations, the general shape of the planet, locations of satellites, etc.

CHARTING PLANETARY MOTIONS

1. You will need a detailed star chart (such as SC1 found in Appendix 4) and a plastic measuring strip. See Exercise 5 for information on making and using the plastic strip.
2. Plot the positions of the planets with respect to the stars on the chart once or twice weekly throughout the semester or quarter. Be sure to date each recorded position.
3. Estimate and record the brightness of each planet compared to the brighter stars for each observation.
4. It is fairly simple to follow a planet's motion with a series of photographs using any camera on which the shutter may be left open for several minutes. See Exercise 8 *Astronomical Imaging* for specific suggestions.

DEEP SKY OBJECTS

1. Set up and align the telescope and start the clock drive if one is available. Accurate alignment will be especially helpful in keeping some of the fainter objects in the field of view.
2. Record the date, location, and general weather and sky conditions for your observations.
3. Record the time and estimate the altitude and azimuth for each object observed.
4. Locate the object with the finder telescope (accurate alignment of the finder with the main telescope is important here). Double stars should be fairly easy to spot, as most will appear double in the finder itself. Nebulous objects are often visible in the finder as small faint blurs.
5. View each object with as high powered an eyepiece as practical and sketch what you see. (Note that lower powers are better for some objects, as high powers tend to "wash them out" or magnify them so much that the field of view is too small.) A wide-angle Erfle eyepiece is often good for larger clusters. Record the focal length or magnification of your eyepiece for each observation made.
6. Be sure to note where appropriate:
   (a) The star colors, relative brightnesses, and orientations.
   (b) The shape of the cluster or nebulous object, and the brightness distribution within it.
   (c) If you know the size of the field of view of your eyepiece, estimate the separations of the stars or the sizes of the clusters and nebulae.
7. Variable Stars. Magnitudes of stars can be estimated by comparing a star to others in the same telescope field of view. Note those stars that appear a little brighter and those that appear a little fainter. Determine the magnitudes of these comparison stars to estimate the magnitude of the star being studied. Good finder charts are essential and planetarium software is very helpful as well. Below are four variable stars to try. Locate finder charts from the American Association of Variable Star Observers at http://www.aavso.org/observing/charts/searchhelp.shtml. Type the star name into the "Pick a star" box and request a chart.
   (a) AM LEO. This is an eclipsing binary of the type W Ursae Majoris, meaning that the period is very short and significant brightness changes can be seen in just a few hours.
   (b) U PEG. This is another W Ursae Majoris star.
   (c) SY CAS. This is a Cepheid variable with a period of about 4 days. You should be able to note changes on any two consecutive nights.
   (d) SX VIR. This is a semi-regular star. It displays large changes in brightness over a period of about 100 days and, thus, makes a good observing target for a semester-long project. If you have access to telescope and a CCD camera then the brightness measurements are much simpler. See Exercise 8.

Table 2-1. Objects to Observe

Name[1]	Object[2]	Constellation	Right Ascension (2000)	Declination (2000)	Comments[3]
All Seasons					
Mizar	Double	Ursa Major	13h 24m	+54°56′	ξ UMa, white-white, 12″.
AlcoR	Op Double[4]	Ursa Major	13h 25m	+55°00′	80 UMa, with Mizar at low power.
Polaris	Double	Ursa Minor	02h 31m	+89°15′	α UMi, yellow-white, 19″.
Spring					
M44, Beehive	Open	Cancer	08h 40m	+19°59′	Wide, but rich cluster visible with naked eye. Use binoculars or lowest power telescope eyepiece.
Algieba, γ Leo	Double	Leo	10h 20m	+19°51′	Both components yellow & quite bright, 4″.
Porrima, γ Vir	Double	Virgo	12h 42m	−01°27′	Yellow-white stars of similar brightness, 6″.
Cor Caroli, α Cvn	Double	Canes Venatici	12h 56m	+38°19′	Both blue, 20″.
M51, Whirlpool Galaxy	Gal	Canes Venatici	13h 30m	+47°12′	Face-on spiral.
M13	Glob	Hercules	16h 42m	+36°28′	Best globular cluster in the sky. Can be seen with naked eye in dark skies. Easy with binoculars.
Summer					
Graffias, β Sco	Double	Scorpius	16h 05m	−19°48′	White-blue, 13.7″, Marked brightness difference.
M4	Glob	Scorpius	16h 24m	−26°32′	Loose, sprawling globular. Easy to find; less than 1.5° W of Antares. Compare with M13.
M6, Butterfly Cluster	Open	Scorpius	17h 40m	−32°13′	Good binocular object in tail of Scorpion. Appears better than most open clusters in low power telescope view.
M7	Open	Scorpius	17h 54m	−39°49′	Just S and E of M6. Easy naked-eye object and better suited to binocular observing than telescope viewing.
M8, Lagoon Nebula	Neb	Sagittarius	18h 04m	−24°23′	Easily visible to naked eye; great binocular object that can be washed away to just the open cluster in telescope view.
M17, Swan Nebula	Neb	Sagittarius	18h 21m	−16°11′	Bright patch of nebulosity. Offers dramatic view through small telescope.
M20, Trifid Nebula	Neb	Sagittarius	18h 07m	−23°92′	Good telescopic nebula about 1.5° NNW of Lagoon Nebula, forming a nice pair.

[1]Star names may vary from book to book.
[2]Double = double stars; Gal = galaxy; Glob = globular cluster; Open = open cluster; Neb = nebula.
[3]Separations of double stars are given in seconds of arc (″).
[4]Alcor and Mizar is a naked-eye pair of double stars separated by 12 arc minutes. Alco and Mizar may or may not be a physical binary system.

Continued.

Table 2-1. Objects to Observe—cont'd

Summer—cont'd					
Double Double, ϵ Lyr	Double	Lyra	18h 44m	+39°40′	Two pairs separated by 208′. Pairs are easily split by binoculars, but high power in telescope required to split each member into two stars.
M57, Ring Nebula	Neb	Lyra	18h 54m	+33°02′	Spetacular example of planetary nebula. Find using a higher power eyepiece than used for clusters and observe with averted vision.
Albireo, β Cyg	Double	Cygnus	19h 31m	+27°43′	Yellow-blue, 35." Great double star with dramatic color difference between the two.
M27, Dumbbell Nebula	Neb	Vulpecula	20h 00m	+22°43′	Bright planetary; compare with M57.
Autumn					
γ Del	Double	Delphinus	20h 29m	−00°01′	Yellow or yellow-white; relatively faint.
ξ Aqr	Double	Aquarius	22h 29m	−00°01′	Both yellowish; not terribly distinct brightness difference between the two.
M15	Glob	Pegasus	21h 30m	+12°10′	Bright cluster on Equuleus border; easy binocular object.
M31, Andromeda Galaxy	Gal	Andromeda	00h 43m	+41°16′	Impressive naked-eye, binocular or telescope object. With telescope use range of magnifications.
Almak, γ And	Double	Andromeda	02h 04m	+42°20′	Beautiful yellow-blue color and brightness contrast. (Blue much fainter than yellow one.)
h and χ Persei, Double cluster	Open	Perseus	02h 20m	+57°08′	Fine pair of clusters lying near back of Cassiopeia and best seen with binoculars.
Winter					
M45, Pleiades	Open	Taurus	03h 47m	+24°07′	Naked-eye cluster; good binocular and decent low--magnification telescope viewing.
M37	Open	Auriga	05h 52m	+32°33′	Outstanding open cluster; shows well in binoculars and is resolved into beautiful splash of stars in a telescope.
M42, Orion Nebula	Neb	Orion	05h 35m	−00°27′	Twisted knot of nebulosity easily visible to naked eye and jumps out in telescope eyepiece. Often regarded as finest object in the heavens.
Trapezium, θ Ori	Cluster	Orion	05h 35m	−05°23′	Four stars at the heart of the great Orion Nebula. Larger scopes show an additional pair of stars.
M35	Open	Gemini	06h 09m	+24°20′	Sparkling cluster visible to naked eye off the feet of the twins. Fine telescope object.
M41	Open	Canis Major	06h 47m	−20°44′	Bright cluster that shows easily with binoculars and fills patch of sky just about right for most low power telescope views.

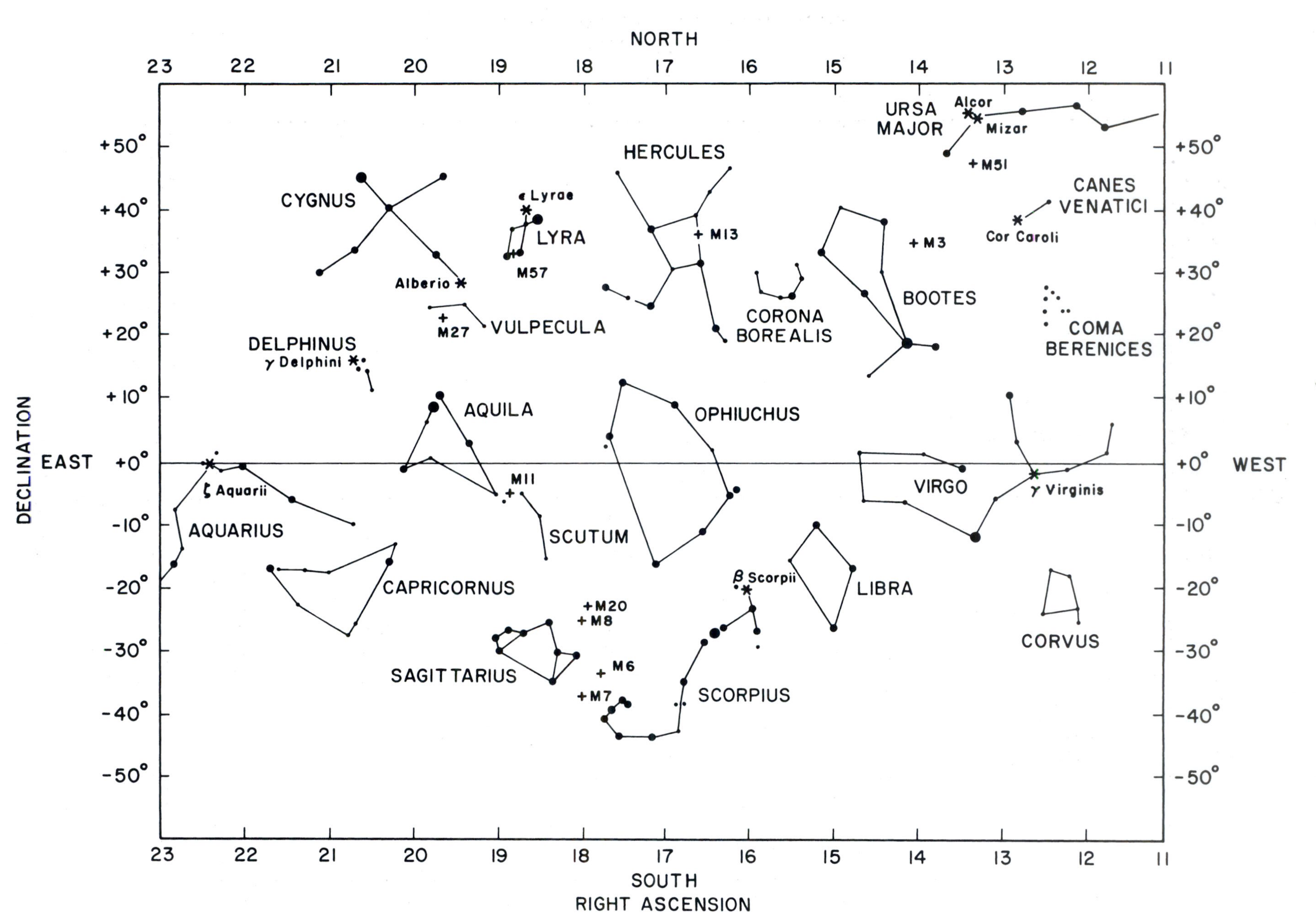

Figure 2-5. Finder chart for summer and fall objects.

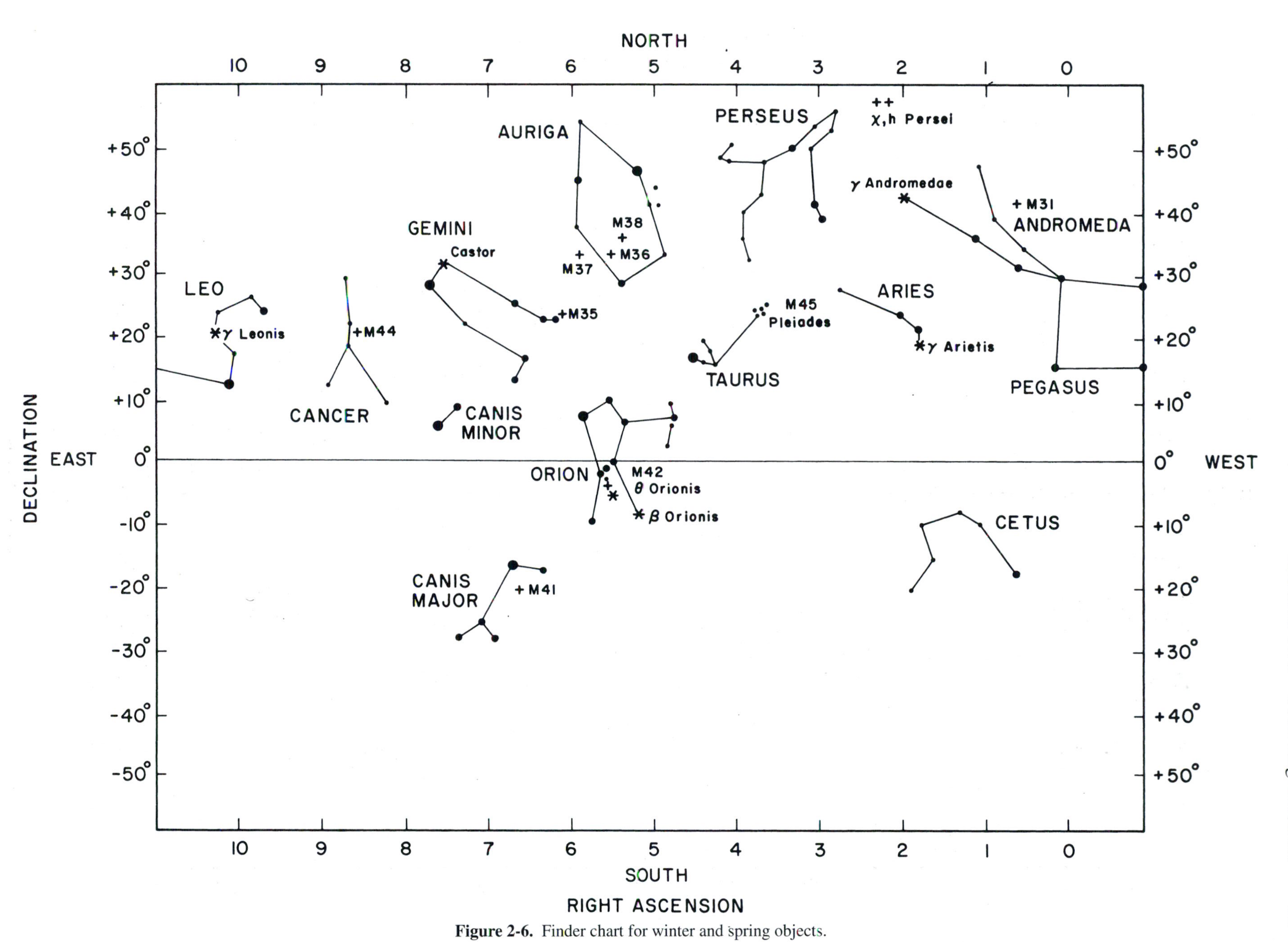

Figure 2-6. Finder chart for winter and spring objects.

Exercise 2. Observing Exercises

LUNAR OBSERVATION DATA SHEET

Object used for setting **RA** circle (RA = right ascension, h = hour, m = minute, DEC = declination.)

Name ______________________ **RA** ______ h ______ m

Time ______ h ______ m (AM/PM)

Moon's **RA** ______ h ______ m **DEC** ______ o ______

LOW MAGNIFICATION VIEW

Crater ______________________

Ray crater ______________________

Mountain range ______________________

Mare ______________________

Mountain peak ______________________

Valley ______________________

Other ______________________

Eyepiece focal length ______________________ mm

Magnification ______________________ ×

Label compass direction

HIGH MAGNIFICATION VIEW

Feature name ______________________

Description ______________________

____

____

____

____

____

Eyepiece focal length ______________________ mm

Magnification ______________________ ×

Draw a circle on the low power drawing to show the field of the high power drawing.

Constellation ______________________ Estimated age ______ d

Exercise 2. Observing Exercises

SOLAR OBSERVATION DATA SHEET

LOW MAGNIFICATION VIEW

Date of observation ____________________

Time __________ (AM/PM) UT __________

Eyepiece focal length ____________________ mm

Magnification ____________________ ×

Declination ____________________

Right ascension (from SC1) ____________________

Zurich sunspot number R = ____________________

(g = ______, f = ______).

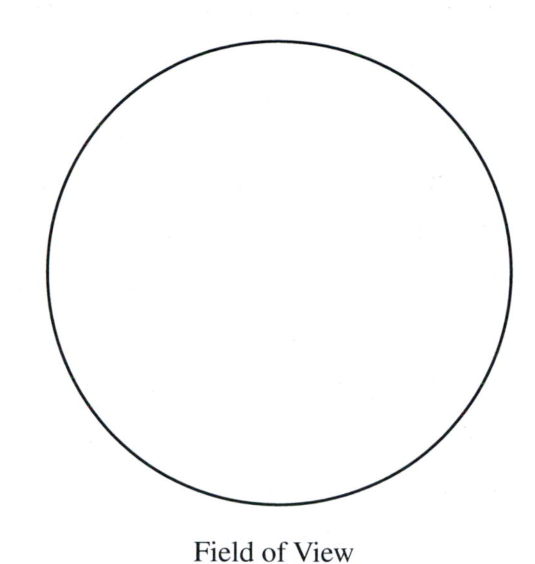

Field of View

Label compass directions in both drawings.

HIGH MAGNIFICATION VIEW

Eyepiece focal length ____________________ mm

Magnification ____________________ ×

Size of the sunspot or sunspot group __________ km

Label the umbra and penumbra of the sunspot or sunspot group.

Sky conditions ____________________
(i.e., clear, cloudy, hazy, windy, seeing, etc.)

Telescope type and no. ____________________

Telescope focal length ____________________ mm

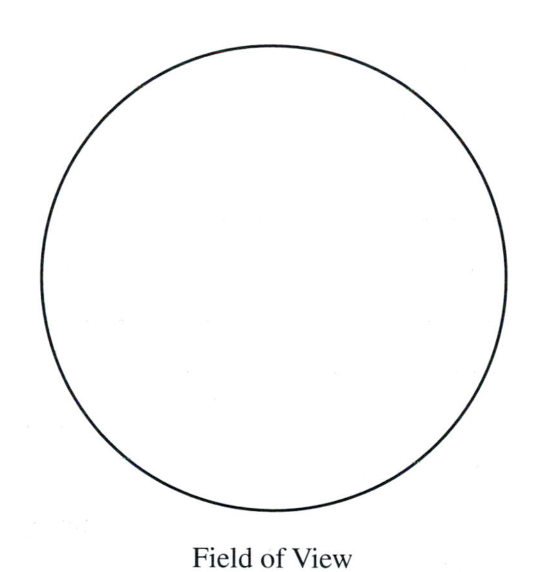

Field of View

Note: Do all drawing in pencil while at the telescope. Do not attempt to work from memory and do not alter this document after leaving the telescope. Draw only what you actually see, and render it as accurately as you can.

Exercise 2. Observing Exercises

PLANETARY OBSERVATION DATA SHEET

Object used to set **RA** circle (RA = right ascension, h = hour, m = minute, DEC = declination.)

Name ____________________ **RA** ______ h ______ m

Time ______ h ______ m (AM/PM)

Planet ______________ RA ______ h ______ m **DEC** ______ o ______

LOW MAGNIFICATION VIEW

Eyepiece focal length ____________________ mm

Magnification ____________________ ×

Use segments to label compass direction.
Note: The circle represents the entire visible field of view, not the size of the object.

HIGH MAGNIFICATION VIEW

Eyepiece focal length ____________________ mm

Magnification ____________________ ×

Description ____________________

____

____

____

____

____

Label any visible moons.

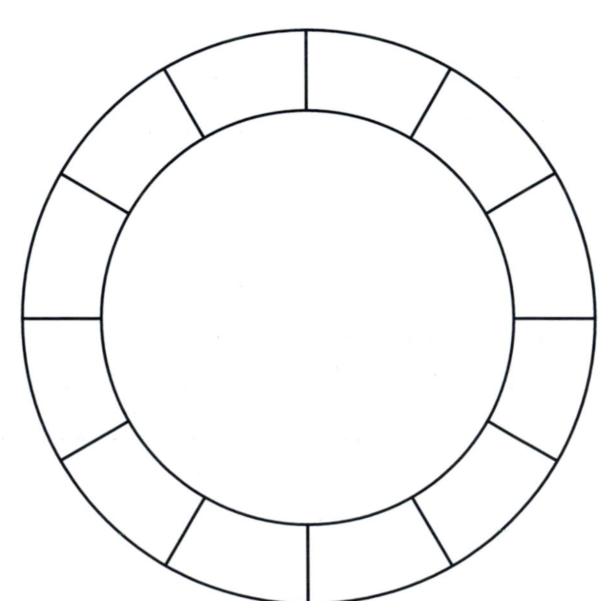

Use segments to label compass directions.

Exercise 2. Observing Exercises

DEEP-SKY OBSERVATION SHEET

Circle setting star __________ **RA:** __________ **DEC:** __________

LOW MAGNIFICATION VIEW

Object Number: __________ **RA:** __________ **DEC:** __________

Name: ________________ Type: ________________ Constellation: ________________

Time __________ (AM/PM) UT: __________ **ST:** __________

Eyepiece focal length ______________________ mm

Magnification ______________________ ×

Description ______________________

____

____

____

____

____

Label compass directions

HIGH MAGNIFICATION VIEW

Object Number: __________ **RA:** __________ **DEC:** __________

Name: ________________ Type: ________________ Constellation: ________________

Time __________ (AM/PM) UT: __________ **ST:** __________

Eyepiece focal length ______________________ mm

Magnification ______________________ ×

Description ______________________

____

____

____

____

____

Label compass directions

Discussion Questions

1. How many different types of lunar features do you see? Can you suggest some of their relative ages or possible origins?

2. Do you see more lunar craters near the edge or near the terminator? Can you suggest several reasons why?

3. Would full moon be a good time to observe craters? Why or why not? Would it be a good time to observe diffuse nebulae? Why or why not?

4. How do the spots move across the sun's disc over a period of a few days?

5. Is the color of the sun's disc uniform? If not, explain.

6. How did the appearance of sunspots you observed change over a time period of a few days? Is the fraction of the sun's disc traversed by a sunspot in a given time interval always the same as the sunspot crosses the disc? Explain.

7. Is the period of rotation the same for sunspots of different latitudes? Can you explain why or why not?

8. Do you notice any motion of sunspots relative to each other in a group? If so, explain.

9. In what general direction does each planet appear to move with respect to the stars? Is the direction of motion constant? If not, discuss any variation.

10. Estimate each planet's angular velocity with respect to the stars in degrees/week, degrees/month, or whatever time unit seems most appropriate. How does this velocity compare from one planet to another? Is it constant for each planet?

11. Describe the differences you observe between open star clusters and globular star clusters.

Exercise 3

Star Charts and Catalogues

Purpose and Processes

The purpose of this exercise is to become familiar with some of the more common star charts, catalogues, starfinders, and timetables. The processes stressed in this exercise include:

Identifying Variables
Predicting
Controlling Variables
Using Logic
Inferring

References

The following are star charts, finders, timetables, and catalogues particularly suited for naked eye observation or for use with a small telescope:

1. *SC1* and *SC2* (see Appendix 4) or similar Constellation Charts
2. *Sky Gazer's Almanac*
3. A Planisphere or Star Finder
4. *The Astronomical Almanac* (current year)
5. *Norton's 2000.0 Star Atlas and Reference Book*
6. *Sky Catalogue 2000.0 Volumes 1 and 2*

Introduction

Star charts and catalogues have been compiled since astronomical events and observations were first recorded. Today there are many devices on the market for locating stars, but here are some of the ones you're most likely to run into (information about how to order them is given in Appendix 3 *Equipment Notes*).

Since this exercise describes the use of several different types of star charts and observing aids, be sure to consult your instructor to determine which parts to do.

SC1 AND SC2 OR SIMILAR CONSTELLATION CHARTS

Most constellation charts are designed to be used outdoors to represent the sky and aid in acctually finding the desired objects. They give right ascensions and declinations, indicate approximate magnitudes, and depict the general pattern the stars in a constellation appear to make.

SKY GAZER'S ALMANAC

This almanac is a graphic plot of local rising, setting and transit times of planets, the sun, the moon, and other bright objects for each date of the year. It also indicates the brighter meteor showers and the phases, apogee, and perigee of the moon.

PLANISPHERES AND STAR FINDERS

These devices are also designed for use outdoors but usually contain less detail than star charts. They have the advantage of adjusting for your date and time to aid in orienting a star chart with respect to your local horizon.

THE ASTRONOMICAL ALMANAC

This reference, published yearly, lists daily positions for the planets, moon, bright asteroids, and the sun; gives data on planetary satellites, eclipses, sunrise, sunset, moonrise, and moonset; lists coordinates for the brighter stars; and gives general data on the solar system. It is often used at the telescope as well as for general reference purposes.

NORTON'S 2000.0 STAR ATLAS AND REFERENCE BOOK

This combination catalogue and atlas is designed for the beginning observer of the sky. It contains detailed constellation charts for telescopic observing and lists data on objects of interest for the naked eye and small telescopes.

SKY CATALOGUE 2000.0 VOLUMES 1 AND 2

This set lists the general star data for stars of magnitude 8.0 and brighter (Vol. 1) and for double stars, variable stars, and nonstellar objects (Vol. 2). The books are available in both paper and hard back from Sky Publishing.

Procedure

CONSTELLATION CHARTS

Two charts, *SC1* and *SC2,* are in Appendix 4 at the back of this manual. *SC1* represents the equatorial region of the sky and *SC2* the north polar region.

1. First look at *SC1.* The heavy line across the center of the chart denotes the *celestial equator.* The wavy line intersecting it represents the apparent path of the sun during the course of a year (called the *ecliptic*). Along the side of the map are numbers running from 0 at the equator to ±60° at the top and bottom. These are *declination* (dec) markings and are analogous to latitude on the earth. They stop at 60° rather than going on to 90° to avoid distortion of the map near the poles. The coordinate going across the top and bottom of the map is called *right ascension* (RA). Since the entire sky goes over our heads once a day, it is measured more conveniently in hours than degrees. 24 hrs = 360° so that 1h = 15°.
2. Locate the following objects and record their approximate right ascensions and declinations. Tell whether each is a star or constellation.
   - Arcturus
   - Aries
   - Orion
   - Sagittarius
   - Sirius

3. The small dates along the sun's path or ecliptic tell when we "see" the sun in that position in the sky. Find and record the coordinates of the sun for today. What constellation is the sun in?
4. At midnight the sun is halfway (or 12 hours) around the sky from our north-south overhead line (called the *meridian*). Thus, stars located 12h from the sun will be on the meridian at midnight. Give the RA of the meridian for midnight tonight and list several stars and/or constellations in that area. (Note: Star charts are set up for standard time. If you are presently observing daylight savings time it will be easiest if you do this exercise for standard time, and convert to daylight savings time as a last step if necessary for local observing.)
5. Since we see the stars moving from east to west during the course of an evening, a star 1 hour WEST of the meridian at midnight should have crossed the meridian one hour before, at 11:00 P.M. Note that on this chart RIGHT IS WEST AND LEFT IS EAST as you look down on it. This is because it is made to be held over your head as you face south. Using standard time, give the RA of the meridian at 11:00 P.M., and name a star or constellation that will be there then. Repeat for 8:00 P.M. and 2:00 A.M. Check your 8:00 P.M. results with the dates along the bottom of the map.
6. This process may also be used in reverse to find the date a given star will be in a good observing position. Follow the example below.

   Example

   On what date will the star Vega be on the meridian at midnight, and when at 10:00 P.M.?

   1. Find Vega: RA = $18\frac{1}{2}$h.
   2. At midnight the sun is 12h away from the meridian, so $RA_{sun} = 18\frac{1}{2} - 12h = 6\frac{1}{2}h$.
   3. Look at the ecliptic at RA = $6\frac{1}{2}$h. The date is June 30.
   4. Vega will be on the meridian at midnight on June 30.
   5. Since the stars all rise and set 2 hours EARLIER each month, it will be on the meridian at 10:00 P.M. one month later, or on July 30.

   On what date will Arcturus be on the meridian at midnight? At 2:00 A.M.? Show your work!
7. Now look at SC2. It shows the north polar region of the sky, and Polaris is at the center. The concentric circles show declination, and right ascension lines run outward like the spokes of a wheel. The best way to use this chart outdoors is to hold it over your head, find the Big Dipper or Cassiopeia, and rotate the chart until it resembles the sky when facing north. Note that each right ascension line has a date with it. This is the date that the stars are on the meridian ABOVE the pole at about 8:00 P.M. local standard time.

   On what date will Caph (in Cassiopeia) be on the meridian at midnight? Alcor and Mizar (in Ursa Major) at 10:00 P.M.? Show your work.

THE ASTRONOMICAL ALMANAC

1. Listings in the *Almanac* are given for universal time, or the time in Greenwich, England, to avoid local time zone confusion. (The longitude of Greenwich = 0°.) Universal time is also broadcast on WWV shortwave radio for navigators and astronomers.

   To convert from local to Greenwich time you need to know your longitude. For example, if you are at approximately 90° W longitude, you are 6 time zones west of Greenwich. Therefore, it is always 6 hours LATER in Greenwich than at your location. For example, 10:00 P.M. (local standard time) + 6 hours = 4:00 A.M. TOMORROW in Greenwich. Thus, for 10:00 P.M. here, you will want to look under 4:00 P.M. on tomorrow's date for the moon's listings. For most purposes the planets move little enough from hour to hour that only daily listings are given.

   Record your longitude and determine what universal time corresponds to 10:00 P.M. local standard time.
2. Locate and record the right ascension and declination of the moon and five naked-eye planets (Mercury, Venus, Mars, Jupiter, and Saturn) for 10:00 P.M. tonight. Watch the signs on declination, as they are not given for each listing in the table.

3. Plot the positions of the moon and planets on your star chart. Which constellation is each in?
4. Which planets will be visible at 10:00 P.M. tonight? In what general part of the sky will they be seen? Remember that for a northern midlatitude observing location you can see approximately 6 hours on either side of your meridian at the equator, slightly more at northern declinations, and slightly less at southern declinations.

PLANISPHERES OR STAR FINDERS

The directions for using most planispheres or star finders are usually given on the back of these commerical devices. (See Exercise 1 *Visual Astronomy* for instructions for how to construct the one found in Appendix 4 of this book.) It is usually faster to determine where things are and what part of the sky you are looking at with one of these devices. However, they are not as accurate or detailed as your star charts and the constellations are often easier to recognize on the larger charts.

1. Set up the planisphere or starfinder for today at 11:00 P.M. standard time. Name several stars and constellations that will be on the meridian. Compare these with your results for #6 in *Constellation Charts* on page 35.
2. On what date will Arcturus be on the meridian at midnight? At 2:00 A.M.? Compare these with your results for #7 in *Constellation Charts* on page 35. When will Caph be on the meridian at midnight? Alcor and Mizar at 10:00 P.M.? Compare to #7 in *Constellation Charts* on page 35.
3. On what date will each of the following be found on the meridian at the times indicated?
   (a) Andromeda at 10:00 P.M.
   (b) Andromeda at midnight.
   (c) Orion at 8:00 P.M.
   (d) Sirius (in Canis Major) at 10:00 P.M.

SKY GAZER'S ALMANAC

The almanac is set up for a latitude of 40° N and a longitude in the center of a time zone. For many observers this will be close enough to your location that corrections need not be used. Check with your instructor; if corrections are necessary, they are given in the explanatory notes that come with the almanac.

This timetable contains more information than most people ever need. We are usually concerned with the rising, setting and transit times of the moon and planets. Detailed information on other uses is in the explanatory notes that come with the almanac.

1. Figure 3-1 shows a portion of a recent almanac in case a current one is not available. Note that the MONTH and DATE are given along the lefthand side and the local TIME across the top and bottom. The complex of diagonal and curved lines on the chart represent various events such as sunrise and sunset; and the planet's rising, setting, and transit times. To find what time a given event happens on a specific date, find the proper event line and trace it to the horizontal line of the date. Then read the time from their intersection point. For example: Saturn transits at midnight (standard time) on September 29.
2. Determine each of the following:
   (a) When does Venus set on February 15?
   (b) What time does Saturn rise on May 3?
   (c) What time does Mercury set on April 26? Will this be before or after astronomical twilight?
   (d) The best time to view the superior planets is when they are high in the sky near midnight. This is called "the transit" of a planet. Will any of the superior planets be in a good viewing position during the last half of the year 2009? If so, which ones?

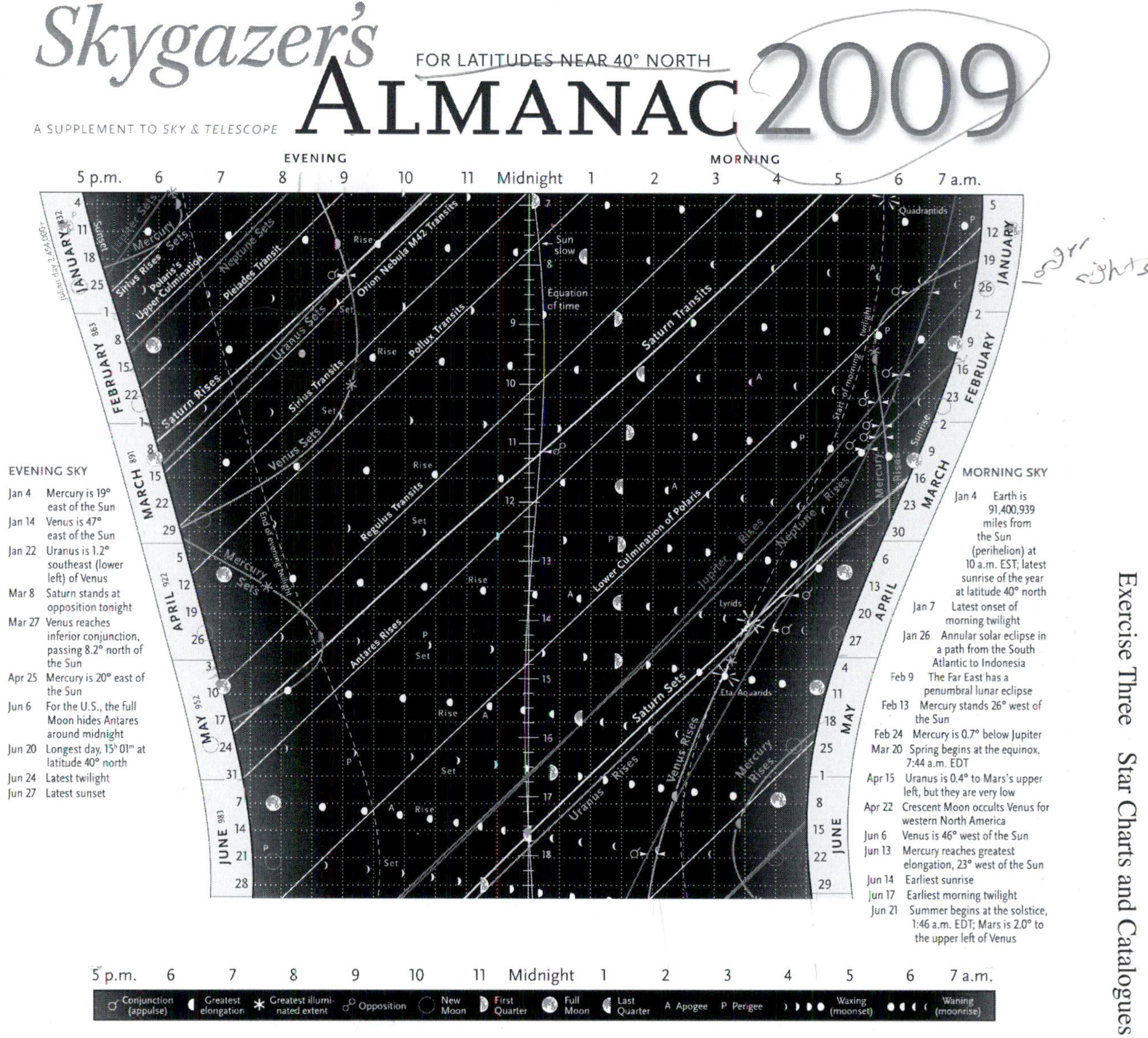

Figure 3-1. A portion of the *SkyGazer's Almanac 2009*. (The *SkyGazer's Almanac* is updated annually by Sky Publishing Corporation and is published each year in the January issue of *Sky & Telescope*. It can also be ordered at http://www.skyandtelescope.com. Reprinted by permission.)

Discussion Questions

1. Do the planets seem to fall in any pattern when plotted on your star chart (in Appendix 4)? If so, what might be the significance of such a pattern?

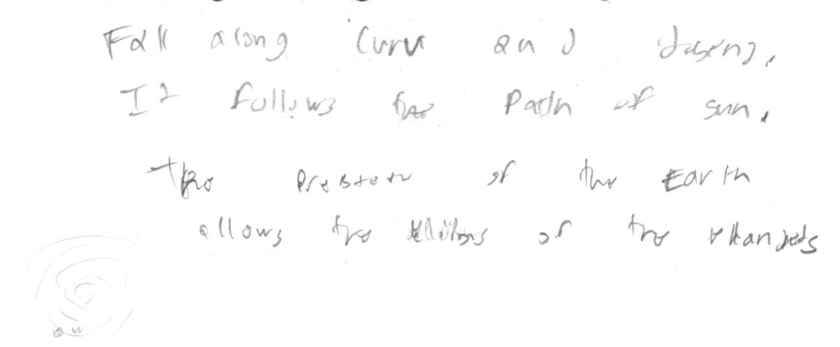

2. We know that the stars rise and set about 2 hours earlier each successive month. Use the *Sky Gazers Almanac* to get the setting time for Mars or Venus for two dates separated by a month. Do they follow the pattern of the stars? If not, why not?

3. (a) What are the relative advantages and disadvantages of using the charts and other devices?

   (b) What is the major source of error in each device?

Exercise 4

Atmospheric Extinction of Starlight

Purpose and Processes

The purpose of this exercise is to study the effect the atmosphere has on the measured signal from stars, leading to a better understanding of how data from stars must be handled.

Using Numbers
Using Logic
Inferring
Interpreting Data

Introduction

The fact that we see the glow of the daytime sky is an indication that the atmosphere scatters starlight. When we say, "My, the sky is pretty today", we might instead say "The scattered light from the sun is particularly attractive this afternoon." When a star, or other celestial object, is close to the horizon, the light from that object passes through more atmosphere than when the object is higher in the sky. See Figure 4-1. If the atmosphere is removing light as the light passes through it, then we should see objects get brighter as they rise and dimmer as they set. If this effect is occurring it is apparent that the calibration of astronomical images is trickier than it otherwise would be. If the atmospheric extinction is not properly accounted for then we might deduce that an object has variable luminosity when, in fact, it does not. In this exercise you will examine data taken with a CCD camera and a small telescope. The data set consists of photometry of 6 stars of varying color, with brightness measures made as the stars rise. You will be able to determine if the stars get noticeably brighter as they rise and if the color of the star appears to play any role[1].

[1] Special thanks to Forrest Bishop doing the initial analysis of the data used here.

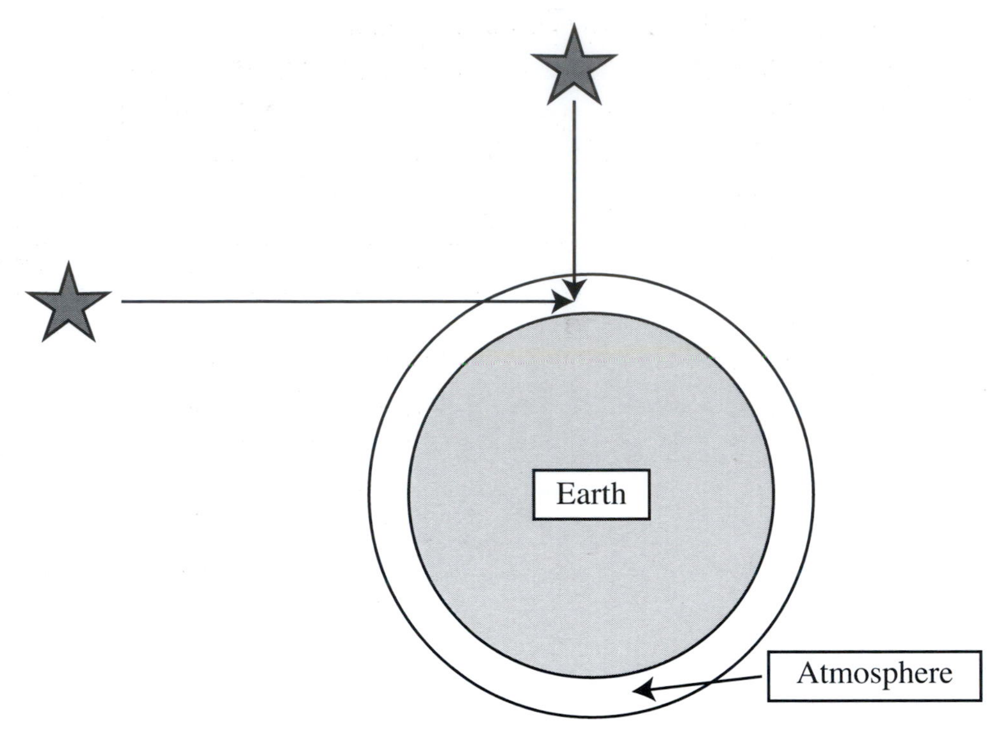

Figure 4-1. Light from a star that is overhead travels a shorter path though the atmosphere than light from a star near the horizon.

Procedure

1. Table 4-1 provides the measured signal from six different stars as a function of altitude. The altitude is the angle between the star and the horizon. The signal is measured in ADU, or analog-to-digital units. See Exercise 8 for an introduction to how CCD cameras are used to make images, how the signal from a star is determined and for instructions on how to take data for this exercise yourself. Graph the measured signal as a function of altitude for each of the six stars.
2. Draw a best-fit line through the data for each of your graphs. Note that we have no expectation that measured signal should be linearly related to altitude. Over a small enough range of altitude, however, the linear fit can still yield a reasonable measure of how quickly the signal is changing. Determine the slope and intercept for each of the best-fit lines you drew. Record the values in Table 4-2.
3. The slope of your best-fit line is a measure of how rapidly the stellar signal is changing but it must be normalized. A change of 100 ADU is far more significant in a star with a signal of 3000 ADU than in a star with a signal of 50,000 ADU. You can find the normalized slope by dividing each slope you measured by the intercept of the best-fit line. Record these values in Table 4-2.
4. Table 4-3 provides color indices for each of the star. The lower the number the bluer the star. These values are taken from the web version of the BDA catalog (Mermilliod and Paunzen, http://www.univie.ac.at/webda//). Make a graph of normalized slope as a function of color index.
5. Write a short paragraph describing your results. Is there compelling evidence that stars get brighter as they rise? Is there compelling evidence that the color of the star plays a role?

Table 4-1. Measured Star Signal as a Function of Altitude

Altitude (°)	Star A Sig. (ADU)	Star B Sig. (ADU)	Star C Sig. (ADU)	Star D Sig. (ADU)	Star E Sig. (ADU)	Star F Sig. (ADU)
19.0	37640	7703	37400	13750	16040	1247
19.3	38070	7689	38070	13800	16390	1257
19.8	38740	7993	39370	14140	16930	1306
20.1	39320	8033	40210	14360	17050	1300
20.4	39520	8123	39560	14720	17360	1316
20.9	39880	8282	40130	14630	17560	1335
21.2	40340	8304	40460	14930	17590	1343
21.5	40670	8304	41150	15230	17880	1391
22.0	41020	8448	41600	15500	18170	1379
22.3	41680	8633	41810	15560	18300	1428

Table 4-2. Linear Fit Parameters

Star	Measured Slope (ADU/°)	Measured Intercept (ADU)	Normalized Slope (/°)

Table 4-3. Star Color Indices

Star	B-V Color Index
A	1.51
B	0.57
C	1.31
D	0.41
E	0.32
F	0.67

Name ____________________ Date ____________________

Exercise 4. Atmospheric Extinction of Starlight

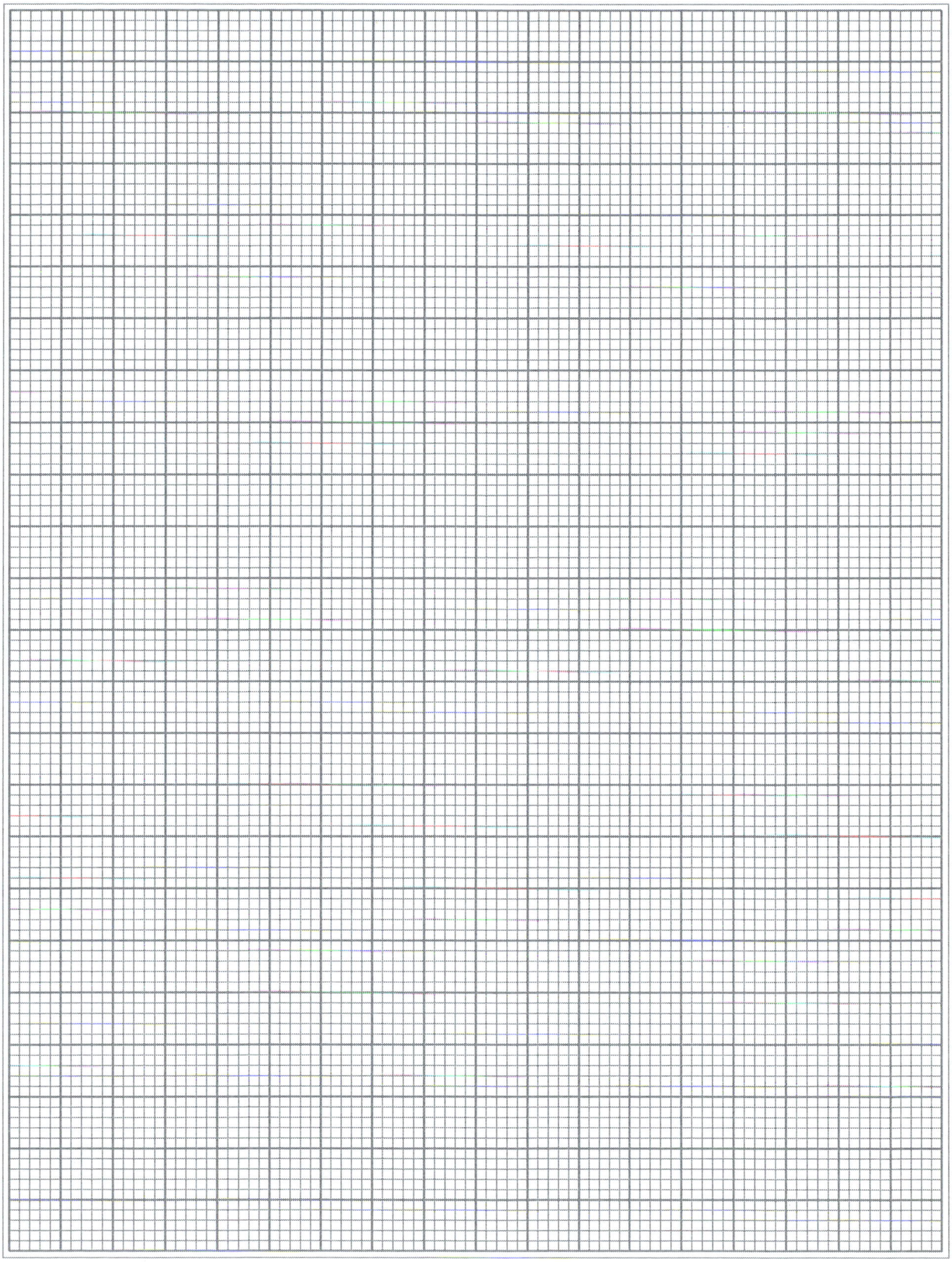

Measured Signal (ADU)

Altitude (°)

Name ______________________________ Date ____________________

Exercise 4. Atmospheric Extinction of Starlight

Measured Signal (ADU)

Altitude (°)

Name ______________________________ Date ______________

Exercise 4. Atmospheric Extinction of Starlight

Measured Signal (ADU)

Altitude (°)

Name ____________________ Date ____________

Exercise 4. Atmospheric Extinction of Starlight

Measured Signal (ADU)

Altitude (°)

Name ______________________ Date ______________

Exercise 4. Atmospheric Extinction of Starlight

Measured Signal (ADU)

Altitude (°)

Name ______________________ Date ______________

Exercise 4. Atmospheric Extinction of Starlight

Normalized Slope (1%)

B-V

Discussion Questions

1. Based on the results you found in this exercise, can you create an argument for why the sky is blue?

2. What might be possible causes of scatter in the data you plotted?

3. If you had continued your signal versus altitude plots to greater altitudes, would you expect the curves to continue in a linear fashion? If not, in what way might they deviate from linear? Explain.

PART II
OBSERVING TOOLS

Exercise 5

Observing with Simple Tools

Purpose and Processes

The purpose of this exercise is to demonstrate that much beginning study of astronomy can be done with simple and inexpensive tools. The processes stressed in this exercise include:

Observing
Inferring
Using Numbers

Introduction

Our ability to model the universe is completely dependent on the quality of the observational data available. In turn, the quality of the data is directly linked to the nature and sophistication of the tools used to observe. For example, Tycho Brahe realized that existing measurements were inadequate for distinguishing between the Ptolemaic and Copernican models. He devoted his life to improving the data available by not only observing regularly for twenty years but also by working continuously to improve his position/angle measuring devices and observing technique. Just a few years later the world of astronomy would be revolutionized by the improved angular resolution and light gathering provided by the telescope.

This exercise includes directions for the use of several simple and inexpensive tools for studying the sky. Your instructor may have you choose one or more of the following sections to complete depending on the equipment and time available for the lab.

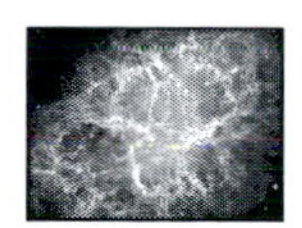

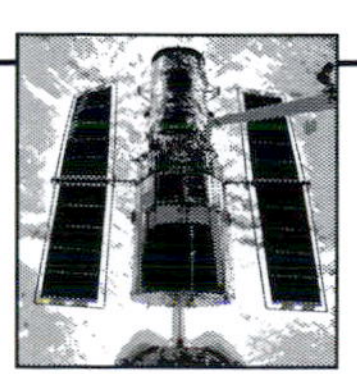

Procedure

STAR COUNTING

Using a cardboard tube we want to determine the number of stars visible to the unaided eye. Consider a sphere whose radius is equal to the length of a cardboard tube (Figure 5-1). The area seen through the tube is equal to the circular cross-sectional area of the tube, given by

$$T = \pi r^2 = \frac{1}{4} \pi d^2$$

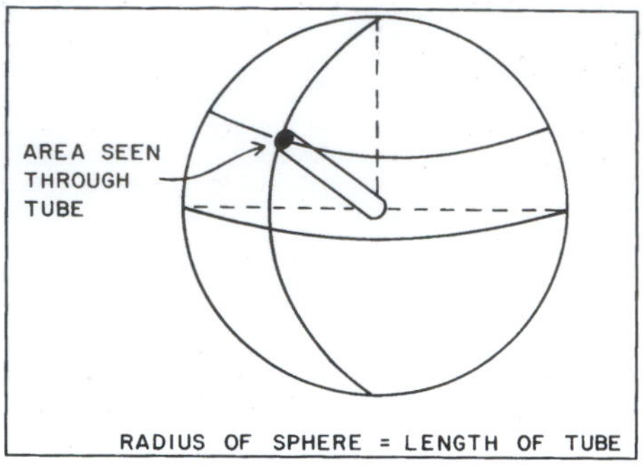

AREA SEEN THROUGH TUBE

RADIUS OF SPHERE = LENGTH OF TUBE

Figure 5-1. Star counting.

where T = area seen through tube
π = 3.1416
r = radius of the tube
d = diameter of the tube.

The surface area of the whole sphere is given by

$$A = 4\pi L^2$$

where A = area of the sphere
L = radius of the sphere = length of the tube.

The ratio of these two areas will equal the ratio of the area of the sky seen through the tube to the area of the whole celestial sphere. If we use the tube to count the stars in N different areas of the sky, the total area of the sky used in counting is

$$C = N\,T = N\left(\tfrac{1}{4}\pi d^2\right)$$

where C = total area used in counting stars.

Now we can set up the ratio

$$\frac{\text{total number of stars counted}}{\text{total number of stars in sky}} = \frac{\text{area of sky used in counting}}{\text{total area of sky}} = \frac{C}{A}$$

or

$$\text{number of stars in the sky} = \frac{A}{C} \times (\text{number of stars counted}).$$

1. Measure the length and diameter of your cardboard tube.
2. Select a number of randomly distributed areas in the night sky (10 to 15 areas should be sufficient). A table of random five digit numbers may be used. Choose 10 to 15 numbers from the table; let the first three digits give the azimuth and the last two digits the altitude of each area of sky.

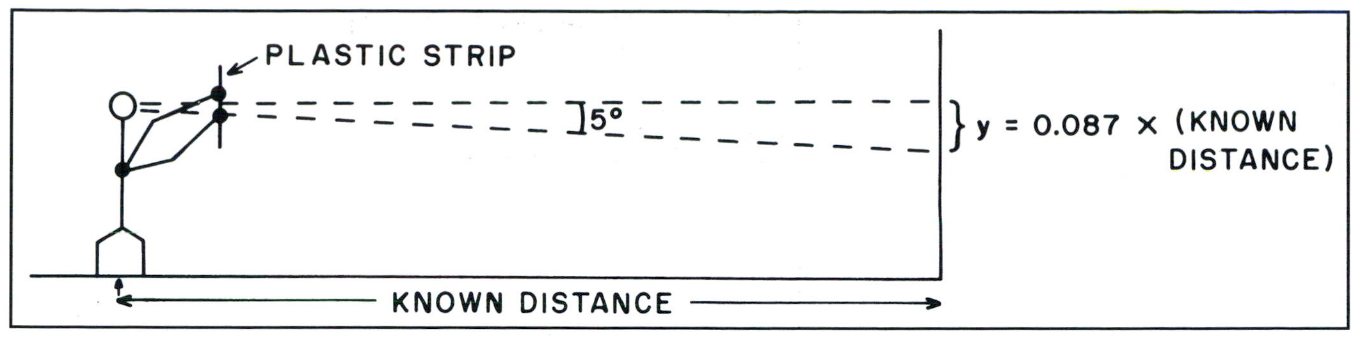

Figure 5-2. Calibration of the plastic strip.

3. Count the number of stars seen through the tube and record the approximate altitude and azimuth for each area.
4. Determine the total number of stars counted, and calculate an estimation of the total number of stars visible to the unaided eye on the celestial sphere. Show all calculations.
5. If possible, do star counts and estimate the total number of stars visible both in the city and at a relatively dark country observing site. Discuss the effects of "light pollution" on astronomical observations in general.

A SIMPLE ANGLE-MEASURING DEVICE

The ability to construct devices to precisely measure angles in the sky occurred before the invention of the telescope. The Dane Tycho Brahe (1561–1626) was recognized for his keen eyesight and for the construction of massive and accurate quadrants. His ability to make and record accurate positional measurement of the planets using the naked eye provided the data from which the German Johannes Kepler (1571–1630) constructed the first empirical astronomical laws.

Angular measurements are still critical to astronomy. For example, being able to measure the angular separation between double stars permits us to calculate their true separation, given that we know their distance. While the device suggested here is very crude, it does permit the measurement of some angles to give you a feeling for the utility of angular measurement.

1. The first step is to make and calibrate your own scale for measuring angles. A transparent plastic strip marked off in 5° intervals is a simple and convenient scale since it can be carried with you easily. The calibration procedure is illustrated in Figure 5-2.

   Stand a known distance from a wall (e.g., 5 m) and hold the plastic strip vertically at arms length between the thumb and index finger of each hand. Line up your right thumb and forefinger with an upper mark on the wall and then your left thumb and forefinger with the lower mark. If the marks on the wall are 0.087 times the distance from the wall apart (.435 m or 43.5 cm for a distance of 5 m), your thumbs will be 5° apart. Hold the strip flat on a table and mark the positions of your thumbs on the strip with a permanent magic marker. Ten- and fifteen-degree intervals can be marked by doubling and tripling the 5° interval, and 1° intervals can be obtained by subdivisions of a 5° interval.
2. Using the plastic strip, measure the angular height of a pole, building, or other object at a number of different distances. Plot your data on a graph and look for a general relationship. Is it linear? If not, can you redefine the variable plotted on one axis and replot the graph so you get a linear relationship?

3. It makes sense that the larger the distance, the smaller the angle for a given object. This relationship can be stated

$$s = r\,\theta$$

where s = linear size
r = distance
θ = angular size.

The relationship originates from a circle (Figure 5-3) where

s = arc length
r = radius
θ = angle

and if the angle θ is less than 15°. The only complication is that θ has to be in units of *radians,* where 1 radian is 57°. This comes from the circle too, as for a WHOLE circle

$$s = \text{circumference} = 2\,\pi\,r$$

and

$$\theta = 360°,$$

so

$$s = 2\pi\,r = r\,\theta.$$

To work, θ has to be equal to 2π. We get 2π radians equal to 360°, or

$$1 \text{ radian} = \frac{360°}{2\pi} \approx 57°.$$

To sum it all up

$$s = r\,\theta\ (\theta \text{ in radians})$$

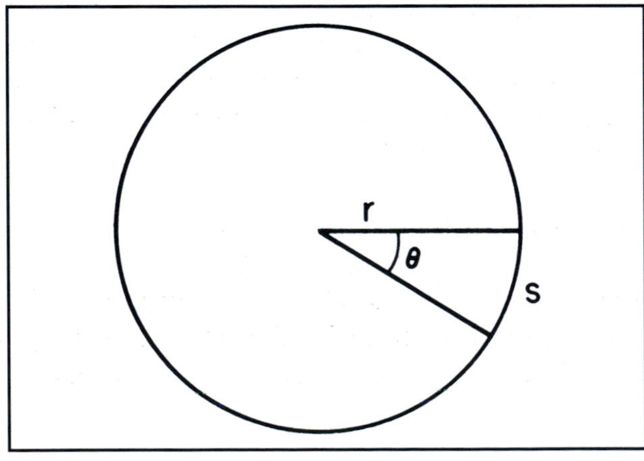

Figure 5-3. s = r θ

or

$$s = r\frac{\theta}{57}\ (\theta \text{ in degrees}).$$

Calculate the height of your building or other object in Part 2 in feet or meters (or in "paces" if you pace off the distances).

4. Determine the height of a classmate at a known distance at the other end of the hall.
5. Practice using your strip to measure the angles between stars or the moon and stars. Can you calculate distances from these measurements?

BUILDING A PINHOLE CAMERA

Today astronomers seldom truly look at the sky. By the end of the 19th century photographic plates began to replace the eye as the principle imaging device in astronomy. Within the last half century electronic imaging devices were developed and attached to cameras. More recently electronically enhanced devices called charge couple devices (CCD) have been developed. Later sections of this book will explore some of these technologies. But photography is still employed by astronomers and it is instructive for beginning students to understand the simple principles of the camera. This section will give instructions for the construction of the simplest imaging device—the pinhole camera.

Since light travels in straight lines and a tiny hole can act as a lens, it is possible to construct a viewing device commonly known as a pinhole camera. Such a device permits beginning students a hands-on way to explore the simple principle of a camera.

1. To construct the device, fasten a piece of black construction paper to the end of a toilet paper tube using a rubber band. Then cut a small piece of waxed paper (or the equivalent) and use another rubber band to cover the other end of the tube. Use the sharp point of a protractor and poke a small hole in the construction paper.

   Turn on a bright light bulb and darken the room. View the bulb by pointing the pinhole end toward the bulb and look at the image of the bulb on the other end. Describe the appearance of the image.
2. A more elaborate version of this device can be constructed using a thin metal sheet and drilling a small hole in it. (A 0.025-inch hole is recommended.) Cut a hole in one side of a large oatmeal box just slightly smaller than the plate. Then cut a small piece of opaque paper and tape one edge of it next to the hole to act as a shutter.

   Obtain some 4″ × 5″ cut film (ISO 400 preferred) and, working totally in the dark, tape a sheet of the cut film inside the box directly opposite the pin hole. Holding the shutter tightly over the pinhole opening, take the box out-of-doors and line it up aimed at some distinct object. Brace and support the box in some fashion so that there is no movement of the "camera" during the exposure. If it is a sunny day a one-second exposure should be sufficient. If it is in the shade or if the day is cloudy, use a 10-second exposure.

   Take the box indoors, unload the film in a darkroom, and follow the developing instructions for the type of film chosen.

BUILDING A SIMPLE REFRACTING TELESCOPE

The Italian Galileo Galilei (1564–1642) was the first to construct a refracting telescope used for astronomical purposes. With it he made several significant discovers that transformed astronomy. The construction of such a telescope is very simple and kits are available from The Astronomical Society of the Pacific. (See Appendix 3 for address.)

Your instructor will assist you in the construction of the telescope. Some of the objects that you may observe to duplicate Galileo's discoveries include:

1. The moon. Observe it. Record its shape and sketch any surface features. Galileo observed the dark flat area that he labeled *maria* or seas. He also observed features that were mountainous and even measured their heights. (See Experiment 16.)
2. Jupiter (if visible). Observe and make a sketch of the positions of its moons over a few-day period.
3. Milky Way. Slowly scan the Milky Way and see if you can distinguish individual stars.
4. Venus (if visible). Observe it and see if you can detect its current shape. Galileo observed it over time and saw that its shape changed in a way that confirmed his belief that it orbited the sun.
5. The sun. DO NOT LOOK AT IT DIRECTLY. Rather try to project it on a sheet of paper and see if you can detect any spots. Galileo was the first to observe sunspots and inferred that the sun rotated.

FIELD OF VIEW OF A TELESCOPE

Knowing how much of the sky you are observing or imaging is useful information for accurate study of the sky. This section describes how you can determine the angular field of view of a telescope.

Since the earth has a rotational period of about twenty-four hours, a star will travel one twenty-fourth of the way around its small circle in one hour of time. If we choose a star that is right on the celestial equator, measure how long it takes to drift across the field of view, and convert this measure of time into an angular measurement, we can determine the field of view of the scope. (If a bright star is at or very near the celestial equator is not visible it is possible to make a mathematical correction to obtain the field of view.)

1. Locate the star to be observed with your telescope. Center the star in the field of view with the telescope drive motor running and adjust the telescope so that the star will traverse the widest part of the field.
2. Set the telescope field just to the west of the star. This puts the star barely out of the field of view such that it will drift through the field when the drive is shut off. Be careful during this maneuver not to change the declination setting, or the star will no longer traverse the center of the field.
3. Turn off the telescope drive. Have a stopwatch ready so you can record accurately the time of transit of the star. Start the watch when the star first appears, and stop it when the star disappears from view.
4. Repeat this procedure at least three times.
5. Determine the angular field of view of the telescope, using the conversions:

1 hour of time = 15 arc degrees
1 minute of time = 15 arc minutes
1 second of time = 15 arc seconds.

Exercise 6

Angular Resolution

Purpose and Processes

The purpose of this exercise is to measure the angular resolution of the human eye and to test whether that resolution is improved by the use of a simple refracting telescope.

Using Numbers
Observing
Designing Experiments
Inferring
Interpreting Data

Introduction

The world of astronomy changed forever when Galileo first trained a telescope on the heavens. His observations of the moons of Jupiter and the phases of Venus provided remarkable support for the Copernican/Keplerian model of the solar system. We often describe a lever as providing mechanical advantage when trying to move or lift an object. In the same way we can think of the telescope as providing an optical advantage when making astronomical observations. This advantage is twofold: the telescope gathers more light for the observation of faint objects and it provides improved angular resolution. Both of these derive from the larger diameter of the telescope when compared to the human eye. In this activity we are interested in angular resolution, or the ability to distinguish two points that are separated by a small angle. The improved angular resolution that the telescope afforded him allowed Galileo to measure the angular size of Venus as a function of phase. The result was stunningly consistent with the solar system model of Kepler. New technology had provided a means for making measurements of fundamental importance—a story that has played out countless times since. We are going to measure the eye's angular resolution and then construct a simple refracting telescope, like Galileo's, to see if our angular resolution is improved.

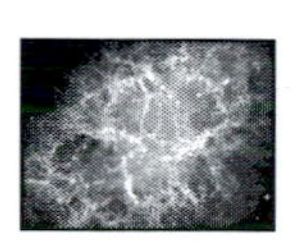

Procedure

ANGULAR RESOLUTION OF THE EYE

1. You know that if you move away from a clock, it gets harder and harder to read the time, because the numbers and lines blur together. This effect is a result of the limited angular resolution of your eyes. Let's measure this resolution. Figure 6-1 is a resolution chart. Stand a few meters from the chart while someone notes which sets of three lines you are able to see as distinct lines and which merge into an indistinct blur.
2. Of the three-line sets that you were able to resolve, measure the line separation in the set having the closest spacing. Likewise, refer to the three-line sets that you were unable to resolve and measure the line separation of the set with the widest spacing. From these measurements, estimate the angular resolution of your eye using $\theta_{res} = s/r$, where s the line separation you are able to resolve and r is the distance from your eye to the resolution chart. Refer to Exercise 5 for more discussion of measuring angles. Record your measured angular resolution in Table 6-1.
3. Move closer to or farther from the resolution chart and repeat your angular resolution measurement, recording the measured value in Table 6-1. Continue to do this until you have about 5 angular resolution measurements.
4. Do your angular resolution measurements look consistent with another? Does there appear to be any trend in the data, such as increased or decreased measured angular resolution with distance? Average your measured values and use that value as the measured angular resolution of your eye. If you are completing this activity with a class, write your value on the board and compare it to others in the class. What would you say is a typical angular resolution for a human?

Table 6-1. Naked Eye Angular Resolution Measurements

r (mm)	s (mm)	θ_{res}

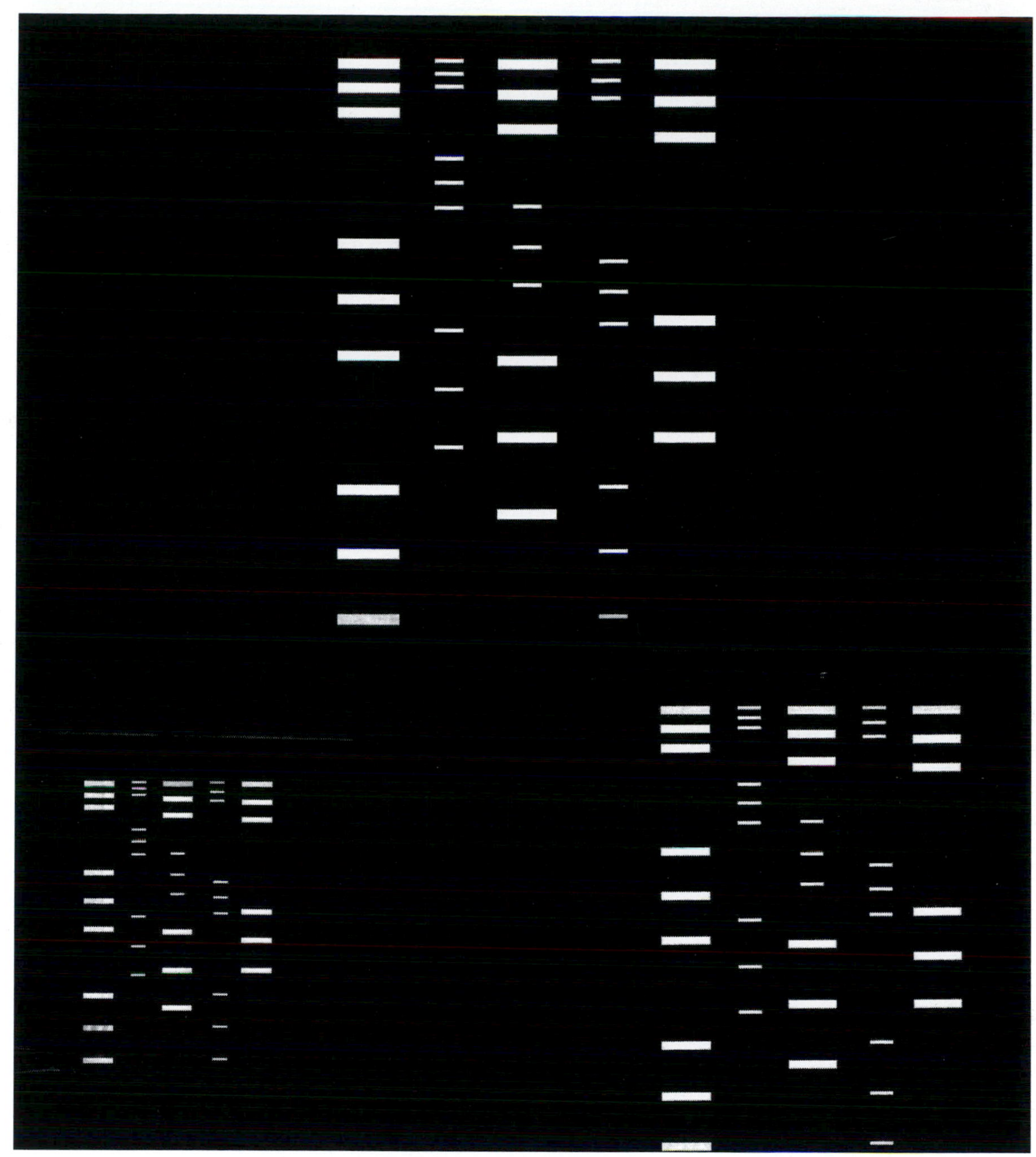

Figure 6-1. Angular resolution chart.

ANGULAR RESOLUTION WITH A TELESCOPE

Now we will build a simple telescope to see if it improves resolution. Our telescope will be a simple refracting telescope, very similar in design to what Galileo used. The lenses bend light as light slows down when it enters the glass.

The light that you are reflecting so that you can be seen is going in all directions. This is also true of the light planets are reflecting and of the light stars are emitting. As the object gets far away from us though, our angular resolution is such that we are not able to see the size of the object and the light rays are almost parallel. The first lens of a telescope takes these parallel rays and focuses the light to a point. The second lens makes the rays parallel again so that you see a brighter distant image. If the object you are viewing is nearby, you see a magnified image.

1. Place two lenses in the lens holders on an optical bench or meter stick so that they can be moved relative to one another. Look through one and slide the other until you get a sharp image of the wall. Use a shorter focal length lens for the eyepiece and a longer focal length lens for the objective. Lenses of, say, 10 cm and 30 cm focal lengths would work well. Make sure you are looking through both lenses when trying form an image.
2. Repeat the steps you completed to measure the angular resolution of your eye looking through the telescope instead. You may need to use larger eye to chart separations. Record your measurements in Table 6-2. Make about five measurements of the angular resolution if possible.
3. Average your angular resolution measurements and record this average as your best estimate of the angular resolution achieved with the telescope. Compare it to the angular resolution you measured for your unaided eye. Divide one value by the other to report by what factor the angular resolution improved or was worsened. If you are completing this activity with a class, write your telescopic angular resolution value on the board and compare it to others in the class.

Table 6-2. Telescopic Angular Resolution Measurements

r (mm)	s (mm)	θ_{res}

Discussion Questions

1. Suppose the headlights of a car are separated by 1.5 meters. At what distance would you be able to see two distinct headlights? Beyond that distance the car would look like a motorcycle.

2. Now try it for the moon. Look up the distance to the moon and estimate the smallest crater or surface feature you can see with your "naked eye".

3. Let's do one more calculation just because we are having such a blast. At closest approach Jupiter is about 6×10^{11} m away from us. How far away from the planets would the moons have had to venture for Galileo to detect them without his telescope? Was the telescope really necessary to detect the? Discuss.

Exercise

7

Image Size—Focal Length Relationship

Purpose and Processes

The purpose of this exercise is to determine the relationship between the size of the image of an extended object and the focal length of the optical system producing it. The processes stressed in this exercise include:

Using Numbers
Interpreting Data
Controlling Variables
Inferring
Predicting

Introduction

The size of an image of a distant object depends only on the focal length of the mirror or lens system producing it. The *image scale* of a telescope is a constant for a particular optical system. It gives the size of the image for each degree of arc the object subtends, often expressed in units of inches per degree, or in millimeters per arc minute.

We wish to find the general relationship between focal length and image size for a one-degree object. Our data consist of several photographs of the moon taken with cameras of various focal lengths. The values are given in Table 7-1.

Whenever we wish to find a general relationship, we are looking for a pattern in the data. One of the

Table 7-1.

Series Number	Focal Length (mm)
1	100
2	205
3	1500
4	3048
5	4570

best ways to find such a regularity is to use a graph. If the data fall along a line passing through the origin, our task is relatively simple, as the relationship may then be expressed in terms of an equation of the form.

$$y = mx$$

or

$$s = mf$$

where s = image size
f = focal length
m = slope of the line
= $(s_2 - s_1)/(f_2 - f_1)$ for any two points on the line.

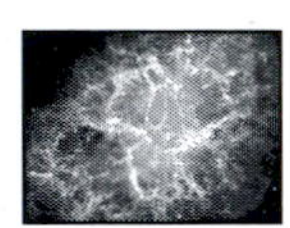

Procedure

1. Measure the size (diameter) of each image in Figure 7-1 to a tenth of a millimeter. Your accuracy will be improved if you make several measurements for each photograph and calculate an average value for each series. For series number five, you will have to measure the size of a crater and compare it to the same crater in series four. Set up a proportion where the ratio of crater sizes is equal to the ratio of image sizes.
2. Enter your data in a table, allowing columns for each step in your calculations. Remember that:
   (a) The prints are enlarged two times, and
   (b) The angular size of the moon is approximately 0.5°.
3. Plot the image scale (the image size for an object of one degree of arc) against the focal length for each of the five series of photographs (labeled graph paper is given.) Fit the best straight line through your data, trying to equalize deviations of the points on either side of the line. Include the point for f = s = 0.
4. Determine the slope of your line and express the image size—focal length relationship in a linear equation.

ADAPTIVE OPTICS

Ever since Tycho Brahe's quest for knowledge of the geometry of the solar system compelled him to devote his life to the refinement of precision angular measurements, our growth in understanding the universe has been fueled by advances in observing and measuring technology. Optical astronomy reached new heights in 1990 with the launch of Hubble Space Telescope. Any image of an astronomical object made from the earth's surface is blurred by turbulence in the earth's atmosphere. This turbulence causes the familiar twinkling of stars. From its orbit far above the bulk of the atmosphere, Hubble's ability to resolve objects is limited only by the size of its mirror. In the years since Hubble's launch, the astronomical community has seen the remarkable development of adaptive optics. In adaptive optics systems, optical sensors measure the atmosphere's turbulence many times per second. A correction is made for this turbulence by using small actuators to change the shape of a telescope's mirror

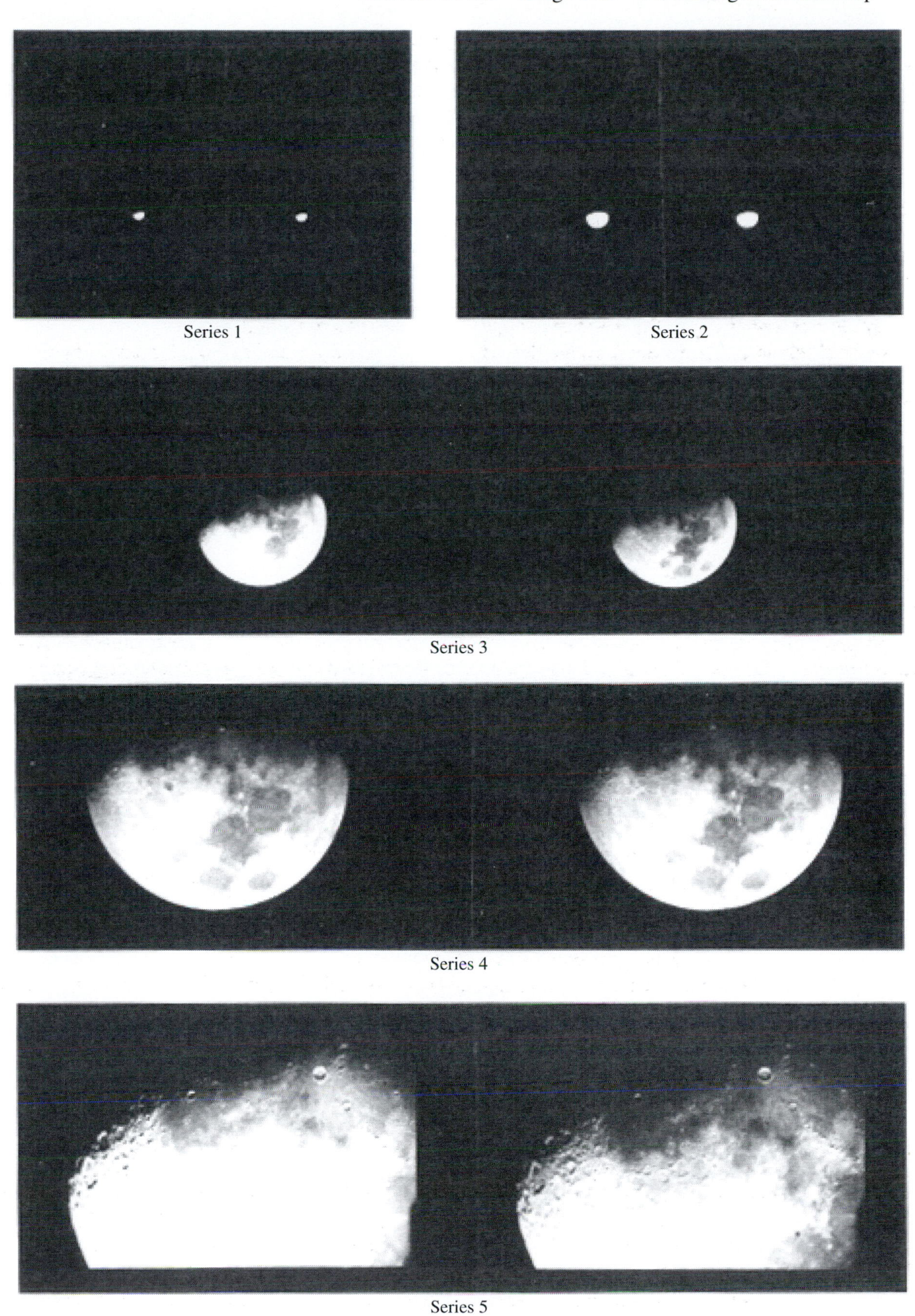

Figure 7-1. Photographs of the moon taken with various lens systems (Courtesy William A. Lane, University of Iowa Observatory)

to correct for the dancing light arriving at the telescope. Typical systems will make adjustments to the mirror surface finer than a tenth of a micron up to 1000 times a second. The use of adaptive optics has been so successful that ground-based telescopes can now rival Hubble Space Telescope in clarity of images. It seems clear that adaptive optics will play an increasingly important role in optical astronomy. Just as Tycho Brahe might have had difficulty imagining the advent of the telescope and its impact on astronomy, many astronomers twenty-five years ago likely couldn't have imagined the development and impact of adaptive optics. What will be the next major advancement in observing technology? That is hard to predict, but surely the advancement will happen and just as surely, it will help us better understand the universe.

Name ____________________ Date ____________________

Exercise 7. Image Size—Focal Length Relationship

Discussion Questions

1. Which of your data points is least accurate and why? How should this be accounted for in fitting a line to the data?

2. What would the image size of the moon be if you photographed it at Cassegrain focus of a 24-inch telescope with a focal length of 9600 mm?

3. An astronomer wants to determine whether or not he or she can photograph Castor and Pollux at the same time on a 5×8-inch plate using a lens system with a 1200 mm focal length. If the two stars are approximately 5° apart, will it be possible to make the photograph?

4. Predict the size of the image of Jupiter if photographed at opposition with a lens having a focal length of 9600 mm. (Assume Jupiter is a disk of 1.43×10^5 km and is 6.29×10^8 km from the earth.)

Exercise 8

Astronomical Imaging

Purpose and Processes

The purpose of this exercise is to make images of the sky using various camera-telescope combinations. The processes stressed in this exercise include:

Using Numbers
Observing
Designing Experiments
Inferring
Controlling Variables
Interpreting Data

References

1. Gordon, G. (2008). *Astrophotography.* Richmond, VA: Willmann-Bell.
2. Arnold, H. J. (1995). *Astrophotography: An introduction.* Cambridge, MA: Sky Publishing.
3. Covington, M. (1999). *Astrophotography for the amateur.* (2nd ed.). London: Cambridge University Press.
4. Howell, S. (2000). *Handbook of CCD Astronomy.* London: Cambridge University Press.
5. Berry R. and Burnell J. (2005). *The Handbook of Astronomical Imaging* (2nd Ed.). Richmond, VA: Willmann-Bell.

Introduction

Photography has contributed extensively toward advancing the field of astronomy. The introduction of photographic techniques in the late 1800s expanded some areas of research and made many new areas possible. For example, about 70 years of work in determining stellar parallaxes had yielded alues for only about 55 stars. In 1904 Schlessinger applied photography to the field and in the subsequent 70 years trigonometric parallaxes were determined photographically for about 6,000 stars. Of course the process of parallax determination of stellar distances has been greatly expanded with the Hipparcos satellite. This satellite has measured parallaxes for about 100,000 stars.

The earliest known astronomical photograph was made in 1840 by Henry Draper (a New York physician) when he completed a successful 20-minute exposure of the moon. In 1845 Fizeau and Foucault obtained the first known photo of the sun, and in 1850

William Bond secured the first stellar photographs of Vega and Castor. In 1875 Sir William Huggins first used photography to capture stellar spectra. In time, improved guiding and photographic techniques made photography an integral part of astronomy.

In the 1970s and 80s a second photographic revolution swept astronomy. This advance was the result of the introduction of electronic cameras based on charge coupled device (CCD) technology. CCDs are solid state light sensors developed in the early seventies and first used to take an astronomical image in 1974. Use of the devices by professional astronomers grew rapidly in the eighties to the point that by the mid 1990s it was difficult to obtain professional quality photographic plates and very few large telescopes were still equipped to use plates. During the 1990s CCD cameras became readily available to the general public as both general use cameras and digital cameras specifically designed for astronomy.

Astronomical photography can be done for pleasure, art or scientific research. A good picture does not necessarily require expensive equipment, but can often depend on the skill, ingenuity, and luck of the photographer. This exercise describes a variety of photographic techniques that the beginning student can use with either a CCD camera or a conventional film camera. Nearly any film camera with a variable shutter speed can be used, but for serious work it is best to have a single-lens-reflex camera (SLR) with a removable lens. Most cameras of this type have a "time" or "bulb" exposure setting and can be attached to a tripod. The film camera techniques involved in this exercise include (1) choice of film types (2) use of SLR cameras, and (3) camera and telescope combinations. Try as many types of photography as time permits. Do not be afraid to try some of your own ideas and do what interests you. Preserve records of your exposures. A sample log sheet is provided. The CCD camera activities introduce fundamental concepts of image processing.

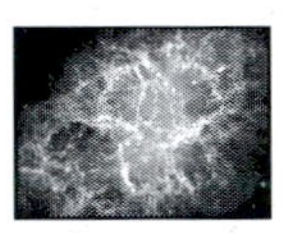

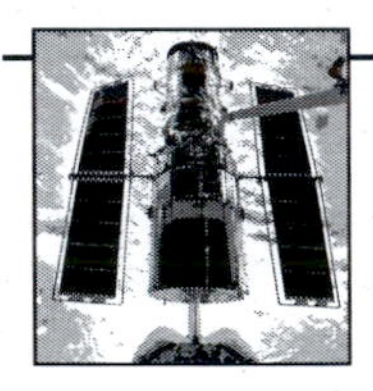

Procedure

CHOICE OF PHOTOGRAPHIC FILMS

An important first step in obtaining good photographs is choosing a film type suitable for your subject matter. Several commercially available films work well for astronomical photography.

1. **Black and White Films.** The current easy availability of rapid film processing of color films has made the choice of black and white film development in schools and colleges less desirable, but there is satisfaction in taking and processing your own film quickly and inexpensively. Black and white films lend themselves to astronomy for this reason. Kodak black and white films currently on the market include a range of ASA or ISO values.

These include Kodak: Plus; Double X; Pro-T Max; and; Ultra Max films. They collectively come in a range of ASA or ISO values from 100 to 3200. Fuji also distributes 400 ASA film. The faster films, regardless of manufacturer, tend to be grainy.

2. **Color Films.** In recent years there has been an expansion of types and ASA films suitable for astronomical work. We shall divide them by whether they are print films or slide films. Only the faster films are recommended.
   - **Print Films.** Kodak sells Ultra Color print film with an ASA of 100 and 400 and Fuji sells Superia Realia 100 (ASA).

- **Slide Films.** They still sell Kodachrome with an ASA of 64, This is apt to be slow for astronomical work. They also sell Ektachrome 100 and Kodak Elite Chrome Extra Color (ASA 100) film. The latter film is now described as being of "extremely fine grained". Kodachrome films favor a color response toward the red-yellow portion of the spectrum and Ektachrome favors the blue-green end.

ASA is an exposure index given by the American Standards Association. Older light meters were calibrated only in terms of ASA exposure indices. Currently film manufacturers are designating film speeds with a two-part ISO number (International Standards Organization). The first number (e.g., ISO 400/28) is the ASA number. Generally one can compensate for the slower speed of some films by lengthening the exposure time. But there is a limit to this reciprocity.[1]

FILM AND DIGITAL CAMERAS

Any camera with a "time" or "bulb" setting can be used for photographing the night sky since it is necessary to keep the shutter open for extended periods of time. However, if you have access to an SLR camera with a good quality lens, you have the advantage of being able to take shorter exposures without loss of image quality. High-speed black and white film as Tri-X or color film such as Ektachrome 200 or 400 are recommended for film cameras. (Ektachrome 200 or 400 films can be processed at ASA 400 or 800, respectively, if requested.) As rule SLR digital cameras have a bulb setting while compact digital cameras are far less likely to have such a setting. Even a compact digital camera that allows exposures up to about 30 seconds might be useful for some applications as noted in the individual activities. If the camera does not have built-in dark noise subtraction capability the user can shoot a dark frame of the same duration as the exposure by fully covering the aperture during a light exposure. The images can then be processed as described in the CCD Imaging section. In this section, unless otherwise noted, it is assumed that an SLR camera with an ordinary 50-55 mm lens is being used, that the camera focus is set at, "infinity", that the aperture is set at its maximum opening, and that a sturdy tripod is available. In addition, use of a cable release will decrease the amount of camera vibration. A suggested technique for long-time exposures is to hold a piece of opaque material over the lens (but not touching it) until the cable release is depressed and locked; then remove the material to begin the exposure. When the picture is completed reverse the process. Longer exposures can be obtained by using the camera "piggybacked" on a telescope as described in the next section.

1. **Constellations.** Excellent constellation slides can be obtained using Ektachrome films and 30-second to 1-minute exposures. A dark sky away from city lights enhances the quality of the slides. The 50-55 mm lens produces image sizes that will encompass all but the largest constellations (such as Draco or Eridanus). Good images can be made with digital cameras as well. Images of a few minutes duration can be obtained with cameras piggybacked on telescopes. These images show fainter stars and regions of nebulosity but the primary asterism of the constellation is less prominent.
2. **Star Trails.** Simply point the camera toward either the polar region or the celestial equator and allow the shutter to remain open for about 30 minutes. Contrast the shape of the star trails and note that Polaris is not precisely at the rotational pole. Contrasting stellar colors show up well on color slides and differences in magnitudes become very obvious. If the sky is extremely dark, longer exposures can be made.

[1]Reciprocity failure refers to the fact that the response of most films is usually not linear with time of exposures of over a few seconds duration. [For high-reciprocity-failure films such as Tmax or color slide or print films, the increase of exposure time by a factor of two or three may allow you to record stars of only one-half magnitude and not a whole magnitude as you might expect.

3. **Comets.** If a bright naked-eye comet is available, excellent photographs can be made with either black and white or color film using exposure times as short as 30 seconds. Photograph the comet on two successive nights to show its comparatively rapid motion among the stars. If possible, it is nice to include part of the horizon to border the photograph and provide a general scale. Again, longer exposures are possible with piggybacked cameras. Fainter, more distant comets can be imaged with CCD or SLR cameras attached directly to the telescope as described later. See the web site for examples.
4. **Aurorae.** Recommended exposure times for aurorae range from 10 seconds to two minutes with Ektachrome films. Excellent auroral pictures with Tri-X film have been made with exposure times as short as five seconds. If possible, try to photograph an aurora against a prominent constellation for added effectiveness.
5. **Meteors.** Meteor photography can be frustrating as meteors are elusive objects. Although five sporadic meteors per hour is a commonly observed rate, you can never be certain in which direction your camera should be pointed. The best chance of success in photographing a meteor trail is during one of the predicted meteor showers. Point the camera about 30–45 degrees away from the radiant point and leave the shutter open for 15 to 45 minutes. If a meteor is seen in the direction the camera is pointing, close the shutter and advance the film to the next exposure. Piggybacked cameras allow longer exposures. Here the meteors will appear as streaks against an the background of point-like stars. When viewing digital images it is best to narrow the display range from faintest to brightest pixel, allowing fainter streaks to be seen.
6. **Bright Artificial Satellites.** Information on passage times of bright satellites can be obtained by using a computer program designed to plot the ground paths of satellites. If possible, pick a time when the satellite will pass through a prominent constellation. Open the aperture to its fullest extent and hold the shutter open for a measured length of time. If the satellite trail and the constellation appear clearly on the negative, make a print and devise a plate scale by using the angular separation of two bright stars. Determine the period of the satellite's orbit.
7. **Planetary Conjunctions and Alignments.** Occasionally two or more planets are seen in the same area of the sky, providing an excellent subject for photographs. If the alignment takes place near the sun (as always is the case for Mercury and Venus), the planets can be photographed just before dawn or right after sunset. By including the horizon in the picture you can determine the inclination of the ecliptic with respect to the horizon. If three planets are visible, see if all three planets fall in the same orbital plane. Five- to twenty second exposures work satisfactorily with color film.

CAMERA-TELESCOPE COMBINATIONS

A camera can be used in several ways with a telescope. Success with any of the methods is dependent on the equipment available and the care and patience exhibited by the photographer.

1. The telescope can be used as a guiding device for an ordinary camera. The camera is attached to the telescope tube and the telescope's equatorial drive keeps the camera pointing at the same region of the sky. This technique is often referred to as "piggybacking" the camera. For extended exposures an illuminated cross hair can be used with the telescope eyepiece to make minor adjustments to telescope guiding. Using a telephoto lens, it is possible to photograph bright asteroids and planets Uranus and Neptune with Tri-X film and exposure times of one to ten minutes. Prominent clusters such as the Beehive, h and χ Persei and Pleiades nicely fill the 35 mm frame with 135 to 200 mm lenses (use 3- to 5-

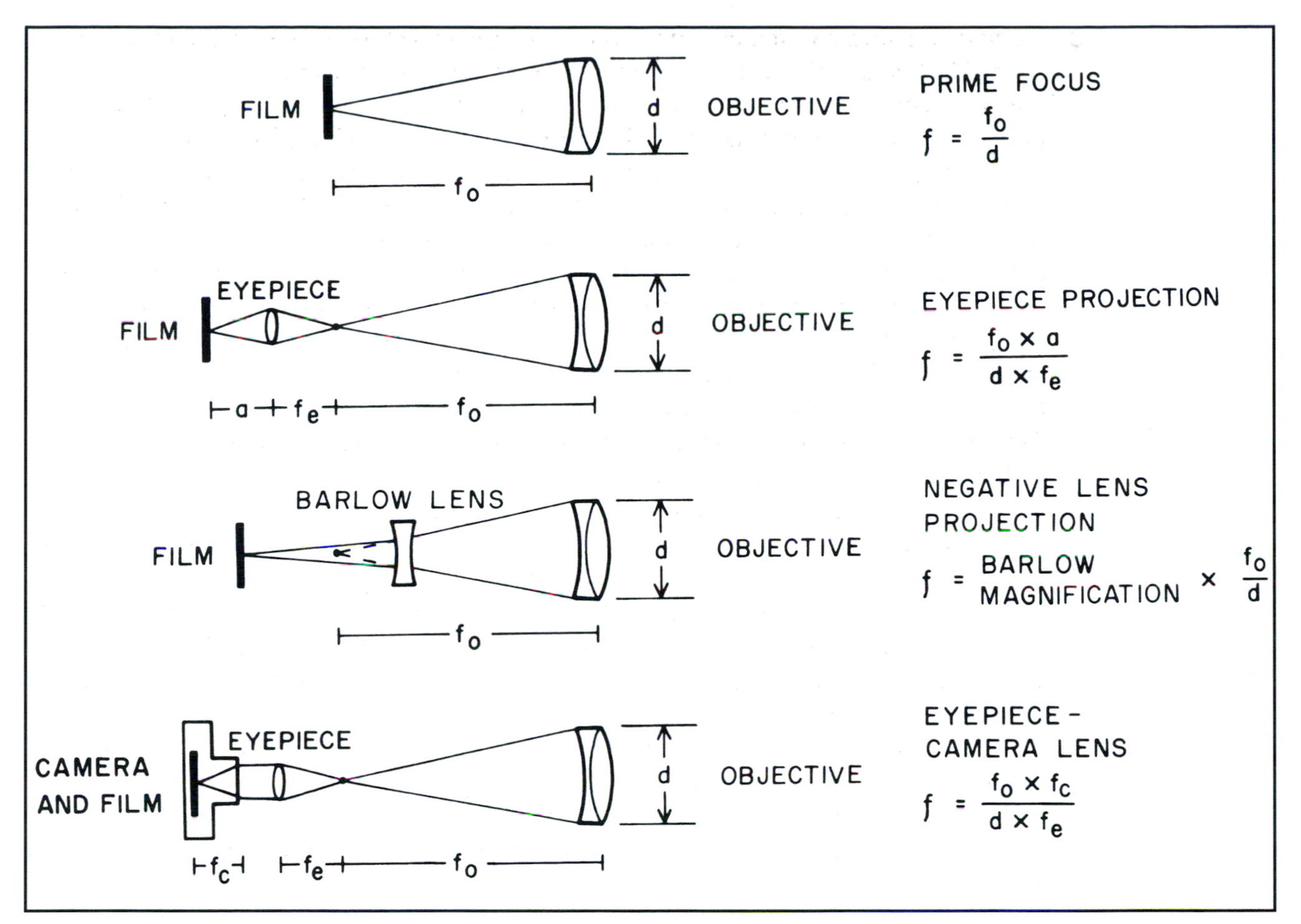

Figure 8-1. Calculating f-ratios for 4 camera-telescope combinations.

minute exposure times). With practice you should be able to locate M 31 and the Ring Nebula on this type of exposure.

2. Four methods can be used to attach an SLR camera to a telescope to take advantage of the telescope's ability to magnify as well as to track celestial objects. An effective f/ratio* can be calculated for each method using Figure 8-1. Suggested exposure times for a variety of celestial objects can be estimated from Table 8-1. It is a good idea to bracket these times with exposure times on either side of the suggested value in order to account for differences in seeing conditions and the cleanliness of the optics of your system.
   - **The Eyepiece-Camera Lens System.** Insert the eyepiece into the telescope, leave the lens on the camera, and mount the camera on a bracket over the eyepiece. (Brackets for this purpose are commercially available or can be constructed locally.) Place a piece of tissue paper over the end of the telescope tube and illuminate it with a 100 watt bulb from about two feet away. Set the camera lens at its smallest aperture. Move the camera in and out on the bracket until the field of view appears evenly illuminated. Remove the paper, focus on a bright astronomical object, open the lens to its largest opening, and set the distance to infinity. Use the telescope eyepiece to focus on the object.

*An f/ratio is the focal length of an optical system divided by the aperture of the lens. The smaller the number (e.g., f/12.8) the "faster" the system is at gathering light. In combination of optical systems as in these descriptions we have to calculate an "effective" f/ratio. These numbers are f _ f/ratio, a _ eyepiece aperature, f_o _ focal length of objective lens, f_e _ focal length of eyepiece lens, d _ diameter of objective lens.

Table 8-1[1]. Astronomical Imaging-Film Photography

Arbitrary Contraction Value												
N	**1**	**2**	**3**	**4**	**5**	**6**	**7**	**8**	**9**	**10**	**11**	**12**
f/ratio	2.0	2.8	4.0	5.6	8	11	16	22	32	45	64	90
ISO	16	32	64	125	250	500	1000					

Exposure Step Change = $N_{f/ratio} - N_{ISO}$

Exposure Time(s)	**Object to be Photographed**
1/4000[2]	Full Sun[3]
1/2000[2]	Venus (greatest elongation); Solar Prominence; 3/4 Solar Eclipse[3]
1/1000	Diamond Ring Effect
1/500	Jupiter
1/250	Full Moon
1/125	Solar Corona; 3/4 Lunar Eclipse
1/60	Mars; Quarter Moon
1/30	Saturn
1/15	Crescent Moon
1/8	
1/4	
1/2	
1	
2	Galilean Satellites
4	
8	
16	Total Lunar Eclipse

[1]Exposure table and procedure developed by Mark Hodges, Owens Valley Radio Observatory.
[2]Not on most cameras.
[3]Exposures are calculated for a Number 4 neutral density filter.

- **Eyepiece Projection.** This is similar to the method given above except no lens is used on the camera. Remove the lens from the camera and mount the camera on the telescope bracket directly over the eyepiece opening. Look through the camera and focus on the object by moving the eyepiece outward slightly from the setting used for visual observation.
- **Prime Focus.** This system produces the sharpest images and is photographically the "fastest" because it uses the fewest optical elements. In order to photograph objects with small reflectors at prime focus an adapter that replaces the camera lens is required. (Most camera stores can provide them.) Such adapters usually come in two parts: an adapter ring that directly replaces the camera lens, and a T-adapter that has an outside diameter equal to that of the telescope eyepiece. The primary telescope mirror will have to be moved forward a short distance in the tube because the image plane of a reflecting telescope is not directly accessible. Some vignetting will occur because of the tube diameter and the size of the secondary mirror. For most astronomical purposes, however, this effect will produce no serious problem. Slide the camera with attached adapter into the eyepiece holder and rebalance the telescope to accommodate the weight of the cam-

era. Focus on the object using the viewfinder of the camera. As in the other combinations used above, the comparaively low level of illumination of celestial objects requires great care in focusing.

- **Negative Lens Projection.** A more common system employed with small reflectors is the negative lens system. Negative lenses (also called Barlow lenses) make it possible to extend the focal length of the system and effectively "pull" the image out of the eyepiece tube without repositioning the primary mirror. As a result this method provides greater magnification. The camera and telescope are set up as in the prime focus system described above with the negative lens used in addition to the adapters. Barlow lenses are usually available from optical supply houses. Solar projection filters also serve as Barlow lenses and this system can be used for sunspot photography. Length of exposure can be determined from published charts (see Table 8-1) or the camera's built-in light meter.

PREDICTION OF EXPOSURE TIMES

The choice of exposure times for various astronomical objects is largely an empirical process involving the f/ratio of the system, the film type used, and the object chosen to be photographed. Table 8-1 presents data that permit a quick estimate of suggested exposure times.

Use Table 8-1 as follows.

1. First calculate the f/ratio of your camera-telescope system by using Figure 8-1.
2. Locate this f/ratio and the ISO of the film to be used on the upper part of Table 8-1. If these two numbers are directly above and below each other, locate the object to be photographed in the lower part of the table and use the listed exposure time. It is good to "bracket" this suggested time by one or two exposures faster or slower to account for possible unusual sky conditions or optics that might not be perfectly clean.
3. If the two values (f/ratio and ISO) do not appear directly above and below each other you must calculate the exposure step change. Find the N values directly above the f/ratio and ISO at the top of Table 8-1. The correction is given by the algebraic difference of these N values where:

$$N_{f/ratio} - N_{ISO} = \text{exposure step change.}$$

4. If the correction is positive, find the exposure time by counting downward from the exposure time listed with the object by a number of steps equal to the difference; if the number is negative, count upward the appropriate number of steps.

   For example, suppose that an f/8 telescope-camera system and Plus-X film are to be used to photograph Jupiter.
   - f/8 gives $N_{f/ratio} = 5$
   - Plus-X film, ISO 125 gives N (ISO) = 4
   - Exposure step ISO change = 5 − 4 = +1.
   - The suggested exposure time is found by counting downward one step from the table value for Jupiter of 1/500s to 1/250s.

CCD CAMERA PHOTOGRAPHY

Imaging with CCD cameras offers the advantages of nearly instantaneous feedback on the imaging process and a digitized data set that is relatively simple to use for making scientifically useful measurements, such as the position and brightness of a given object. In addition, exposure times are significantly shorter than with film since quantum efficiency (fraction of incident photons converted to usable signal by the detector) of the silicon chips is much higher than the quantum efficiency of film. Unlike film, these devices offer a linear response. The measured signal is a linear function of the photon flux striking the detector. Double the number of incident photons and the measured signal doubles, once noise has been removed. One advantage of resulting from this linearity is that it is possible to take several short exposures, summing the signal in them to get the equivalent signal of a single exposure that is the duration equivalent of the sum of the shorter images. For example, the signal from a star will be the same in 6 10- second exposures as it is in a single 60-second exposure, if noise is subtracted appropriately. This ability to sum short exposures greatly eases the requirements on telescope tracking ability.

UNDERSTANDING CCD IMAGES

The CCD camera functions in much the same way as a film camera when attached to the telescope, except that the film is replaced with a two-dimensional array of silicon pixels that serves as an imaging photon detector. When light strikes the chip, electrons are liberated and stored in each pixel until the camera is read and the signal is amplified and digitized. The number of electrons read from a particular pixel depends not only on the amount of light striking the chip but on the efficiency of detection. Electrons are also generated as they escape the silicon without light being present. This dark noise can be reduced by cooling the chip and most cameras have the ability to cool the chip about 40°C below ambient air temperature. Still, to get good photometric data or for the best visual quality images, one must apply calibration frames to images. These take the form of flat fields, bias frames and dark frames. Flat fields are images acquired with the telescope/camera system uniformly illuminated. These are used to remove pixel-to-pixel variation in sensitivity. Most flat fields are taken of the sky under twilight conditions. Bias frames are taken without opening the camera shutter and without collecting charge. They are used to remove the read noise that is present independent of image duration. The camera shutter is also left closed for dark frames but charge is collected for the same amount of time as the light exposure. To reduce noise one can take longer dark exposures and scale them to the correct exposure time. Of these calibration frames, dark frames are often the most important and many camera-operating software packages have a mode that allows dark frames of the correct duration to be taken when the light frame is exposed. Several software packages for image processing exist and many of them are also capable of running select CCD cameras. Adobe *Photoshop* is a popular example of processing software and *CCDSoft* from Software Bisque and *MaxIm CCD* by Cyanogen are examples of software capable of image processing and camera control. *AIP4Win* is a software package that comes with a CCD imaging tutorial book. The package, by authors Berry and Burnell, is available from publisher Willman-Bell. This software allows quick and simple measurements of object positions and signals. We offer these only as examples of what is available and we encourage you to explore the options available to you.

One of the most important aspects of using a CCD camera is understanding image size and image size relative to pixel size. In Exercise 7 you are asked to determine the image size/focal length relationship, but we will present the result here so that we may use it. We can use what we know about the optical properties of telescopes to write:

$$\text{image size (mm)} = \text{telescope focal length (mm)} \times \text{object size (radians)}$$

We rarely measure the angular extent of astronomical objects in radians, instead using arc seconds. Converting to arc seconds gives:

$$\text{image size (mm)} = \frac{\text{focal length (mm)} \times \text{object size (arc sec)}}{2.06 \times 10^5}$$

Since pixel sizes are about 10 microns, we might want to convert the image size to microns:

$$\text{image size (microns)} = \frac{\text{focal length (mm)} \times \text{object size (arc sec)}}{206}$$

Another convenient way to write this same information is in the form that tells us how much sky a pixel of our CCD camera sees:

$$\frac{\text{arc seconds}}{\text{pixel}} = \frac{\text{pixel size (microns)}}{\text{focal length (mm)}} \times 206$$

Multiplying by our array size in pixels yields the field of view of the camera. These calculations offer a good way to estimate how much sky a given camera/telescope combination will see but to make precise measurements you must measure the actual scale of the image. Planetarium software (e.g., *The Sky*) makes this simple. Find two stars in the image and locate them in the planetarium software. The software should provide the angular separation of these stars. Dividing the angular separation by the pixel separation in the image yields the scale of the image, a number that can be directly compared to the value derived above. One can then determine the effective focal length of the system.

QUICK ASTROMETRY

The digital camera is a powerful tool for measuring precise positions of objects on the sky (astrometry). Here are some simple exercises to get you started in using digital imaging for astrometry. These exercises are specifically designed to work with any camera/telescope combination and they are relatively insensitive to image defects such as overexposure.

The Moons of Jupiter

1. Take an image of Jupiter and its moons every ten minutes for about 4 hours. Take more if you have the time.
2. Make a print out of each image. Enlarge the images as necessary.
3. Using the best precision measuring devices you have available, measure the distance of each moon from the planet on each print. Make a table of time and separation for each moon, labeling them moons 1 through 4. If you have software that measures the separation for you, use that.
4. Were you able to see each of the moons move? Plot separation vs. time for all four moons on a single graph. Do you need more data the next night?
5. Identify each of the moons. Use a current chart as necessary.
6. When doing this exercise do you prefer a long focal length telescope or a short one? Why?
7. Explain how you might extend this exercise to measure the mass and density of Jupiter.

The Planets

1. Take images of Mars separated by two hours. Make the exposures deep enough to see background stars even if Mars is overexposed.
2. Make a print out of each image. Enlarge the images as necessary.

3. Using the best precision measuring devices you have available, measure the distance of the planet from a nearby star on each print. Try a few stars. If you have software that measures the separation for you, use that.
4. Were you able to see Mars move against the background of stars? If so, estimate its change in position in arc seconds, arc minutes or degrees.
5. Take an image the next night and repeat the above procedure, comparing Mars' position to its position the previous night.
6. Repeat the entire procedure with an outer planet, Jupiter or Saturn. You may, of course, take the images on the same nights you are observing Mars.
7. Suppose you were to use a focal reducer that shortened the focal length of your telescope by a factor of three. How would it affect your experiment? What changes would you make in procedure to compensate?

For photos of Uranus and Neptune taken over several nights check this activity's web page on the book web site. Can you use the images to determine the relative orientations of Earth and these planets in their orbits at the time the photos were taken?

QUICK PHOTOMETRY

One can take advantage of the speed, ease of data storage and linearity of CCD imagers to make quick and relatively precise photometric measurements. Photometry is the measurement of how bright an object appears to us. Aperture photometry is the simplest form of photometry. Here, all measured signal inside a circular aperture centered on a star (or other object) is summed. An annulus around this inner circle is skipped and the sky background is determined from an annulus outside this "dead" annulus. The sky background is subtracted from each pixel within the signal aperture. If the frame has been dark noise subtracted and flat-fielded the result is a relatively precise measure of the flux of the object being imaged. In principle, one could make these measurements by hand, determining signal pixel by pixel. Software such as *AIP4Win* makes aperture photometry simple.

As we see below, the measured flux of objects will change due to changing air mass and varying observing conditions. These variations must be corrected for in order to search for any intrinsic luminosity variations in an object of interest. It is typical to use a pair of stars within the field as standards for calibration. In a star-rich field, one can simply force the median measured signal of all stars in the frame to be the same in any pair of frames. When this photometric procedure is used, the precision of brightness measurements for faint objects is typically limited by the precision of sky background measurements. The brightest stars in the field have photometric precision limited by frame normalization. In between these extremes photoelectron counting statistics limit the photometric precision. Visit the web site for a typical curve of measured signal uncertainty as a function of signal.

Checking Linearity

1. Pick any region of the sky you like and take an exposure that is approximately 1 second in duration. Subtract appropriate dark frames and use image analysis software to determine how much signal was measured for several stars in the field. Here the signal is measured ADU, or analog to digital units. It is best to avoid a field with a bright star since that star is likely to saturate pixels.
2. Repeat the above measurements with fifteen or twenty different exposure times. Pick a few that are shorter than the 1 second exposure above and several that are well spaced in duration up to as much as a couple of minutes. Doing an experiment such as this it is easy to appreciate the CCDs quick acquisition of data and the data's immediate availability for study.

3. Construct graphs of measured signal versus exposure time for the stars you are studying. You can make a different graph for each star or put the data for all your stars on one graph.
4. If you take several exposures of a single duration you can use the scatter in the measured signal of any star as the uncertainty in the detected signal for that star. Do your graphs appear to show a linear relationship between detected signal and exposure time within uncertainty? That is, can you reasonably draw a line that fits your data? If the CCD camera is a truly linear device you would expect to be able to do this. Report on the linearity of your camera.
5. If you see some of the stars deviating from a linear relationship at one end of the data range or the other (large or small signals) what might be the cause?

Rising Stars

1. Pick a region of sky and track it as it rises. A bright open star cluster will offer you a field with many stars to study but just about any field will work. See Exercise 4.
2. Take a few short dark- subtracted exposures of the field noting the time the exposures were taken. You can adjust the exposure duration to be long enough to get a good measure of the signal from several stars but short enough that no or very few bright stars are saturating pixels. When doing photometry it is critical that you check for pixel saturation! Discard any star image that appears to be even within a few percent of saturation. If you have avoided very bright stars, exposure times in the 2 to 10 second range should be adequate.
3. Repeat the process of step 2 every ten minutes as the star field rises and reaches maximum altitude.
4. Pick a given star and find its average measured signal for each set of a few exposures taken in steps 2 and 3. What is happening to the apparent brightness of the star as it rises? Does the earth's atmosphere affect the star enough to be detectable?
5. Let's develop this measurement further. Use appropriate software or star charts to determine the right ascension and declination of the field of view you have been imaging. Use similar resources to select three stars in your field that offer a range of colors. Pick a relatively blue star, a relatively red star and one somewhere between these two. If you are observing field stars, the Hipparcos satellite web site (http://astro.estec.esa.nl/SA- general/Projects/Hipparcos/hipparcos.html#content) might be a useful resource. If you are observing a cluster, the WEBDA web site (http://obswww.unige.ch/webda/) might prove useful.
6. *(OPTIONAL)* Using whatever tools you have available find the altitude (= height above the horizon) of the imaged field of view at the time of each set of exposures. These altitudes can be calculated from latitude and longitude information. Planetarium software, such as The Sky, will provide the altitude of a star or other object any desired time.
7. If you found the altitude of the stars at the time of each exposure make graphs of measured signal versus altitude. Otherwise, simply graph measured signal versus time. You can make three separate graphs or plot data for all three stars on one graph.
8. Do your measurements suggest that all stars are affected equally by the atmosphere, independent of the color of the star? If not, are bluer or redder stars affected more? Write a paragraph summarizing your results and suggesting further measurements.

Variable Stars

1. Many stars are variable on timescales of just a few hours. The variability of these stars can be measured easily within a single night of observation. Stars for study can be gleaned from the *General Catalog of Variable Stars* or charts prepared by the *American Association of Variable Star Observers.* The web page for this activity will direct you to the web sites associated with these sources. Look for Delta Scuti, Beta Cephei and W Ursa Majoris stars since they have short periods and their variability can be detected within a few hours of observation.

2. Longer period eclipsing binary stars also offer outstanding targets for photometric studies. Algol (β Persei) is a leading example. Eclipse times of the Algol system can be found in many publications such as *Sky and Telescope.* Of course, the sources listed above also list many long period variable stars. If one points a telescope/CCD system at any star-rich region of the sky for a sufficient length of time it is quite likely that an eclipsing system (previously known or not) will be caught in eclipse ingress or egress.

For data to complete this exercise check this activity's web page on the book web site.

Name ______________________________ Date ______________

Exercise 8. Astronomical Imaging—Film Photography

PHOTOGRAPHY LOG SHEET

Film Type ______________________

Roll Number ______________________

Lens System	Frame	Exposure Time	Object
	1		
	2		
	3		
	4		
	5		
	6		
	7		
	8		
	9		
	10		
	11		
	12		
	13		
	14		
	15		
	16		
	17		
	18		
	19		
	20		

Exercise 9

Kirchhoff's Laws and Spectroscopy

Purpose and Processes

The purpose of this exercise is to become familiar with Kirchhoff's Laws and the techniques of laboratory spectroscopy. The processes stressed in this exercise include:

Observing
Formulating Models
Inferring

Introduction

It is important to understand the properties of light because most of an astronomer's information about celestial objects arrives in the form of electromagnetic waves. Astronomers, both in the laboratory and at the telescope, have spent a great deal of time devising techniques to decode the message of light in order to yield as much information as possible. By examining a mere pinpoint of light we can now determine many of the physical properties of the body from which the light originates; properties such as temperature and chemical composition.

One of the astronomer's most useful tools is the spectroscope or spectrograph, which spreads light into its spectrum or rainbow of composite colors. We find that many properties of light can be explained in terms of its behavior as a wave. The *wavelength* of light determines its color: red light is made up of waves almost twice as long as those of violet light. The various colors of light we see are simply waves of different lengths. The order of colors is the same as that seen in a rainbow: red (longer waves) through orange, yellow, green, blue, and violet (shorter waves).

In fact, light (or as it is more generally called *electromagnetic radiation*) comes in many colors we can't see: for example, ultraviolet rays are those just shorter than violet, and infrared rays are those just longer than red (Figure 9-1). Our visible light is just a small part of the whole electromagnetic spectrum, which includes x-rays (with very short wavelengths) to the longest radio waves. The white light we see is just a combination of all the colors of visible light.

An instrument used to form a spectrum from a light source is called a *spectroscope* if the spectrum is viewed visually, or a *spectrograph* if it is recorded on photographic film or plates. The light in either instrument is dispersed or spread out into its colors by a prism or diffraction grating. A slit to make a thin beam of light and one or more lenses are also necessary. A prism spectrograph is schematically represented in Figure 9-2.

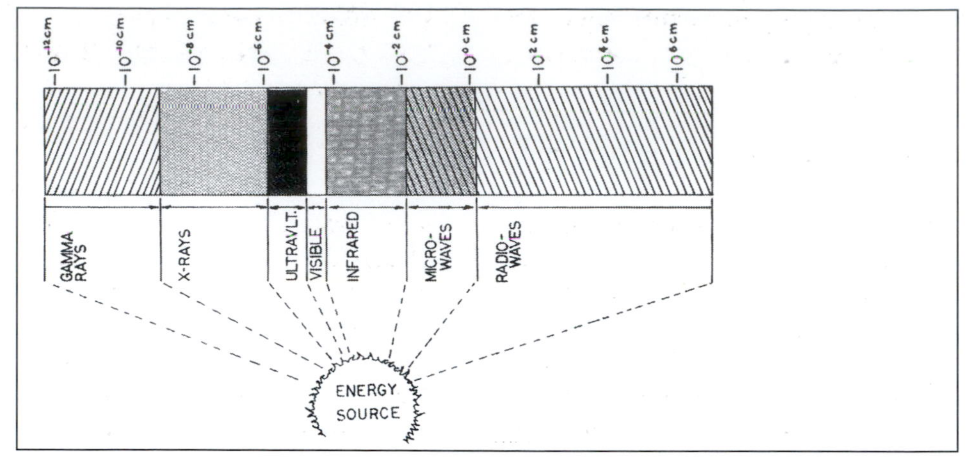

Figure 9-1. The electromagnetic spectrum. (From ASTRONOMY ILLUSTRATED by Bonneau & Smith Kendall/Hunt Publishing Co. 1975.).

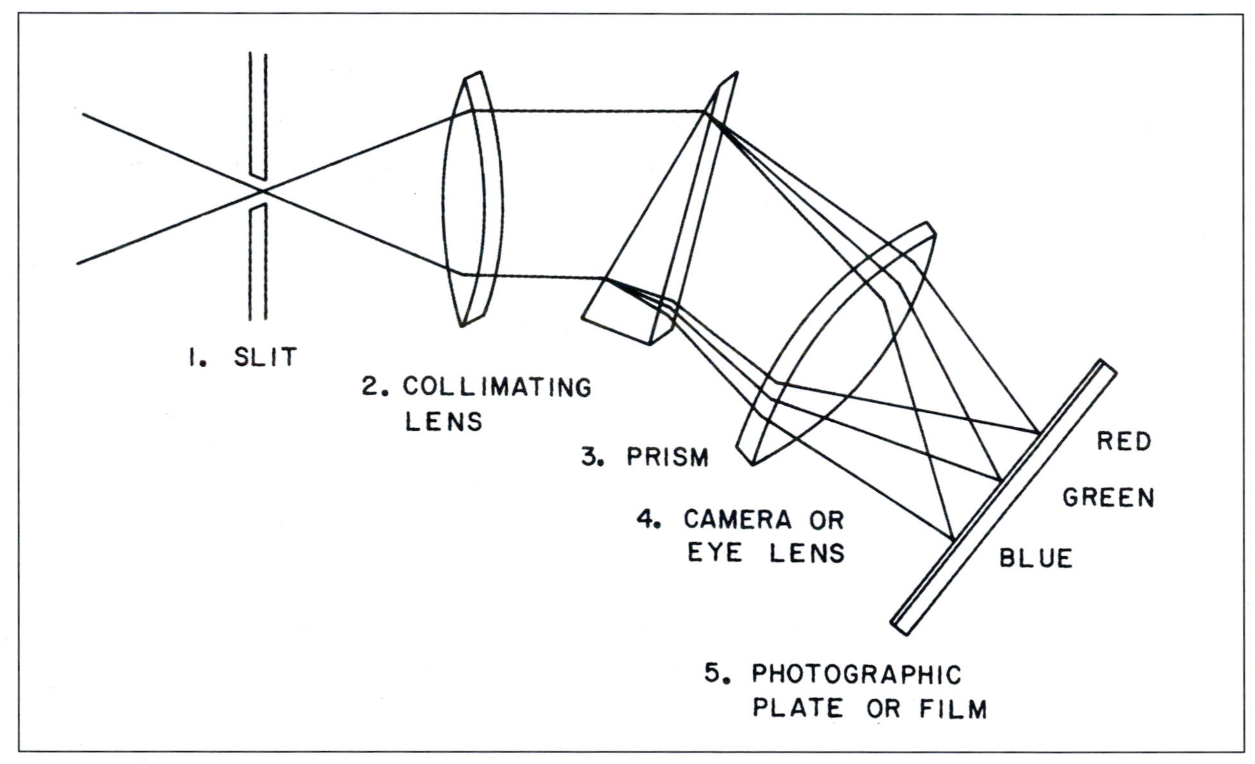

Figure 9-2. A prism spectrograph. 1. The slit is placed at the focal point of the objective; 2. The collimating lens makes the light parallel; 3. The prism disperses the light; 4. The camera lens focuses the light on photographic film; and 5. The film records the spectrum as an image of the slit for each individual color or wavelength.

A prism works by bending the different colors of white light by different amounts as they pass through it. A grating is made by ruling a series of very fine parallel lines (often 20,000-30,000 lines per inch) on a mirror or piece of glass. It produces a spectrum by the more complicated processes of diffraction and interference of the light. The grating has an advantage over the prism in that it has equal dispersion at all wavelengths, while prisms disperse blue light more than red. Replica gratings can now be produced at very low cost.

KIRCHHOFF'S LAWS

In the 1800s, when spectra from various luminous objects were first being studied, it was found that all spectra were not like our customary rainbow: some have bands of color missing, and others have bright bands of color. Gustav Kirchhoff divided spectra into three main types when he proposed explanations for their origins. These explanations are Kirchhoff's Three Laws of Spectral Analysis.

- **FIRST LAW.** A luminous solid, liquid or very dense gas emits light of all wavelengths, thus producing a *continuous spectrum* or *continuum.* This spectrum appears as a continuous change of colors from red to a deep violet (like a rainbow). We should remember that it really extends beyond the red through the infrared and beyond the violet into the ultraviolet regions even though our eyes cannot detect such wavelengths. The continuous spectrum is nearly the same for all substances and varies in total brightness with the temperature of the solid or liquid. The distribution of intensity as a function of color also varies as the temperature is changed. Continuous spectra give almost no indication of chemical composition.
- **SECOND LAW.** A rarefied or low-pressure gas emits light whose spectrum contains bright lines, sometimes superimposed on a faint continuous spectrum. This is a *bright line* or *emission spectrum.* It appears as a set of distinct sharp lines of different colors, separated by darker spaces. The wavelengths of specific lines present are characteristic of the atoms of the gas. Each type of atom or element in the periodic table has its own unique set of lines. If the gas is a combination of several types of atoms then the spectrum will contain lines characteristic of each element. The emission spectrum therefore is of great importance in determining the chemical composition of the gas.
- **THIRD LAW.** If a continuous spectrum from a luminous source passes through a cool tenuous gas, certain wavelength regions will be extracted from the continuum causing dark lines (or absorption lines) to appear. This *absorption spectrum* looks like a continuum with certain sharp lines blotted out. Researchers have found that these lines are at exactly the same positions that would be occupied if the same gas were emitting rather than absorbing. Thus these absorption lines are just as useful as emission lines in identifying the composition of the gas.

The three types of spectra are depicted in Color Plate 7.

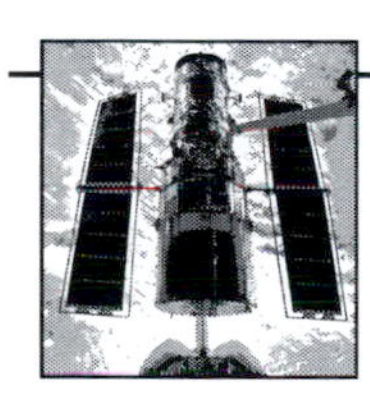

Procedure

OBSERVING WITH A DIFFRACTION GRATING OR SPECTROMETER

You will be supplied with either diffraction gratings or a small hand-held spectrometer. Diffraction gratings are mounted in small slide mounts and should be held up close to the eye. If you look at a bright light source, this device breaks the light into its component spectral wavelengths. The spectrum will appear to the left and right of the central source. If you don't see the spectrum, turn the grating 90 degrees and try again. Remember to look to the right (or left) of the light itself. A spectrometer is a simple device containing a grating *and* a wavelength scale. This device permits estimates of the wavelengths of the light being observed. A very useful small spectrometer was developed with National Science Foundation funds at the Harvard-Smithsonian Center for Astrophysics and is pictured in Figure 9-3.

OBSERVING THE CONTINUOUS SPECTRA

Your instructor will set up a clear light bulb plugged into a variac or similar variable power source. First simply observe the light with the naked eye as your instructor starts with a low power setting. Observe both the level of illumination and the color of the bulb. Then continue observing as the instructor increases the power to the bulb by slow stages. Record both the brightness and the color of the filament.

Your instructor will repeat this experiment and ask you to examine the bulb with a grating or spectrometer. Attempt to determine where the brightest part of the spectrum falls as the

Figure 9-3. Project STAR simple spectrometer.

instructor increases the power to the bulb. This is a physical demonstration of Wien's Law, which relates the peak in the spectral energy to the temperature of the source. Astronomers use this law to determine the temperatures of stars.

IDENTIFICATION OF ELEMENTS FROM THEIR SPECTRA

It is possible to produce emission spectra of many elements by heating a wire dipped in certain chemical salts, or by using gas discharge tubes. Produce as many spectra as possible and view each with a spectroscope. (See data sheet on page 88).

For each trial:

(a) Record the general color of the flame or tube as it appears visually.
(b) Sketch the spectrum indicating both the COLORS and RELATIVE SPACING of the lines. Colored pencils may be used if available.
(c) If the element being observed is unknown, sketch it as above, and identify it by comparison with other spectra you have recorded or with a wall chart of common spectra.

1. Gas Discharge Tubes

   Each tube has been evacuated and then filled with a certain type of gas. When a high voltage is applied across the tube the gas glows and emits light.

   BE CAREFUL IF IT IS NECESSARY TO REMOVE THE GAS TUBES FROM THE TUBE HOLDERS. The use of gloves is often required.

2. Flame Test

   You will be supplied with a platinum test wire and a bunsen burner and several salts such as calcium oxide (CaO), strontium bromide (SrBr), and potassium bromide (KBr).

   It is possible to produce a gas from each of the salts (Ca gas, Sr gas, Or K gas) for a short time (several seconds at most) by dipping the platinum wire into the salt and holding it over the bunsen burner flame. Be sure to dip the wire into hydrochloric acid before dipping into the salt each time. Since the duration of the colored flame is short, it may be helpful to have one person produce the flame while several others view it.

   If a propane torch is used instead of a bunsen burner, observe and record the appearance of its spectrum. Note the wide absorption bands and indicate their approximate positions in your drawing. These absorption bands are called the Swan spectrum and are due to carbon (C_2) in the flame of the torch. They are of special interest to the astronomer as they are often observed (and thus imply the presence of C_2) in the spectra of comets.

COMMON LIGHT SOURCES

1. Use the spectrometer to examine the spectra produced by as many of the following common light sources as possible:
   - An incandescent light bulb
   - Fluorescent lights
   - Street lights and spotlights of various types
   - Matches
   - A candle

   Indicate the type of spectrum each produces (continuous, absorption or emission). If it has lines, sketch them as you did above and identify the elements present.
2. Look at the light from an incandescent source with and without a holmium oxide or similar filter. Note what changes the filter makes in the spectrum's appearance. Use two filters, and again note any change.
3. If you have access to the small spectrometer, take it out of doors and look at sunlight reflected from a cloud or from some smooth surface. Look carefully and you should be able to see the solar spectrum as a generally continuous spectrum crossed with dark lines. These are the absorption lines first observed by Fraunhofer in the early 1800s.

Thoughts from Ann Parsons . . .

Designing More Precise Telescopes

I am an astrophysicist working in the Laboratory for High Energy Astrophysics (LHEA) at NASA/Goddard Space Flight Center. Those of us working in "high energy" astronomy study the X-ray and gamma-ray emission from astronomical sources to deduce what physical processes must be present to create light at such high energies. For example, we can learn a lot about the properties of a black hole by studying the X-ray light emitted by matter as it is drawn in by the black hole's extremely high gravitational field. There are many different ways to do astronomy but the idea is basically the same: we try to determine the nature of an astronomical body by the amount and type of the light it emits. Whether we measure low frequency radio signals or very high frequency gamma-ray emission, the frequency or energy of the electromagnetic radiation we detect tells us a lot about the physical processes occurring at the astronomical source.

While the basic idea is the same, the actual telescopes used to detect this high energy light look very different from the visible light telescopes most people are used to. For example, since Earth's atmosphere absorbs the X-ray and gamma-ray emission before it reaches the ground, telescopes for high energy astronomy must be placed above the atmosphere in either scientific satellites or high altitude balloons. The mirrors in X-ray telescopes look very different than in optical telescopes because mirrors can only reflect X-rays at grazing angles. In fact, gamma-rays and X-rays at energies higher than 10 keV will go right through the mirrors and not be reflected at all. For these higher energy photons, the total collecting area of the telescope is limited to the size of the X-ray detector itself.

Much of my work at NASA is directed toward the design and construction of new X-ray and gamma-ray telescopes that can provide more precise photon energy information from fainter and fainter sources. Technical breakthroughs in X-ray and gamma-ray mirror and detector design can lead to significantly more sensitive instruments that can provide great opportunities for new astronomical discoveries. Although I've been interested in astronomy since I was 10 years old, I never expected to be doing astronomy in quite this way. I've been curious about the universe since childhood, but I also found that I like to work with my hands in the laboratory. My job here at NASA/Goddard gives me the opportunity to balance the study of the universe through analysis of existing satellite gamma-ray data with the hands-on hardware technology development that I enjoy.

Ann Parsons
Staff Scientist
NASA/Goddard Space Flight Center

Name ______________________________ Date ______________

Exercise 9. Kirchhoff's Laws and Spectroscopy

DATA SHEET

SPECTRA OF KNOWN GASES

	Gas	Spectrum	General Color
1.			
2.			
3.			
4.			
5.			
6.			
7.			
8.			
9.			
10.			

SPECTRA OF UNKNOWN GASES AND THEIR IDENTIFICATIONS

	Gas	Spectrum	General Color
1.			
2.			
3.			
4.			

Discussion Questions

1. A faint continuum is often observed along with the bright emission lines from a gas discharge tube. Can you suggest its origin?

2. If an "air" tube is available, compare its spectrum to those of the other gases you observed. Do you note any similarities? Discuss.

3. If spectra were viewed with filters in *Identification of Elements from their Spectra,* what differences did one or two filters make in the spectrum? Can you generalize from your observations?

4. What does Kirchhoff's explanation of the type of spectrum of the sun tell us about the sun's physical nature?

PART III

MEASURING DISTANCE IN THE UNIVERSE

Exercise 10

Angles and Parallax

Purpose and Processes

The purpose of this exercise is to introduce simple techniques for measuring angles, to use these techniques to become familiar with the relationship between distance, angular size, and linear size; and to study the principles of the parallax method of determining distances to celestial objects. The processes stressed in this exercise include:

Using Numbers
Identifying Variables
Controlling Variables
Using Logic

Introduction

We use the relationship between the angular size of a known object and its distance almost automatically when we are driving an automobile. A car which "looks half as big" (subtends an angle half as large as another automobile) is approximately twice as far away because automobiles are roughly the same size. We use the fact that we know the approximate linear sizes of common objects such as cars, buildings, and bicycles to estimate their distances from their apparent angular sizes. The mental conversion of an observed angular size into an estimated distance is done easily by most persons for common objects that are relatively close. However, we would have difficulty in estimating the distance to the Goodyear blimp. We don't see it very often and, if it is high in the sky, there is no common object at the same distance to compare it to. The eye can measure only the ANGULAR separations or angular sizes of objects; it takes the mind to convert to real physical measurements. We want to make this process more systematic and quantitative, and use a simple and practical technique for measuring angles to measure distances and/or linear sizes of various objects on the ground.

One of the basic problems in astronomy is determining distances to celestial objects. By definition, *parallax* is one-half of the angle formed at the celestial body by two intersecting lines drawn from the ends of a baseline. In the case of objects in the solar system, the baseline is a diameter of the earth, and in cases of objects outside the solar system the baseline is the diameter of the earth's orbit. Technically these are called *geocentric* (or *horizontal*) parallax and *heliocentric* (or *annual*) parallax, respectively (Figures 10-1 and 10-2).

If the moon is observed from the two end points of a diameter of the earth it will be seen in two different positions with respect to the background stars. The total amount of apparent displacement (2p) is about two degrees, the exact value depending on the

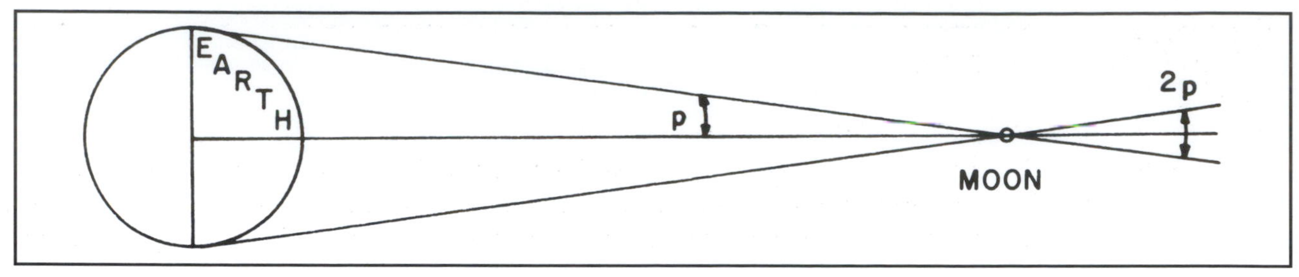

Figure 10-1. Geocentric or horizontal parallax.

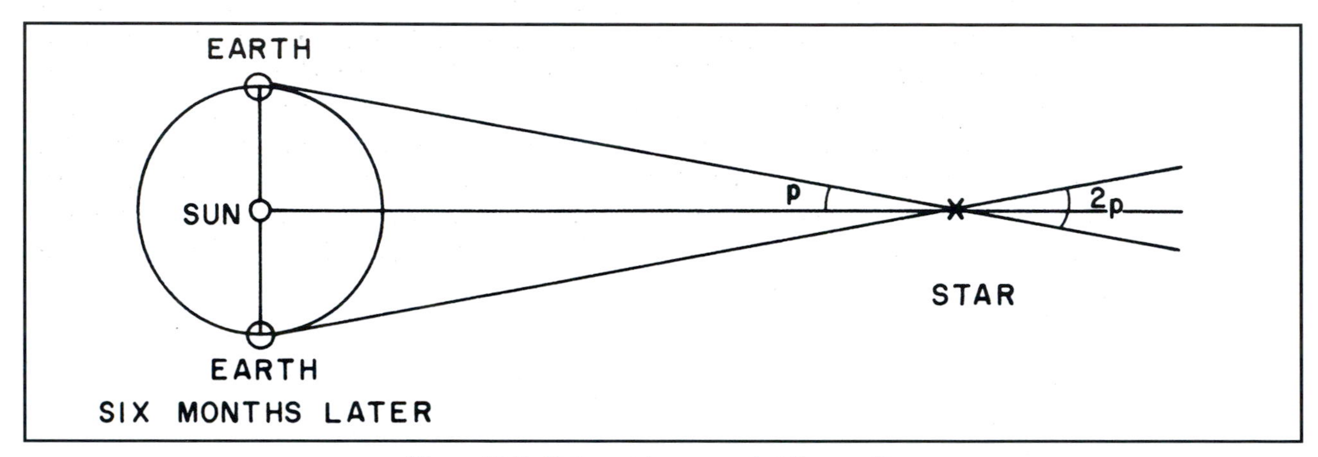

Figure 10-2. Heliocentric or annual stellar parallax.

earth-moon distance. Since parallax is defined as one-half of this displacement (p), the parallax of the moon is about one degree.

A similar shift is seen for all stars during the course of one year. Such small displacements are involved that the measurement of stellar parallax is most commonly done by photographing a field of stars at intervals of six months, and then measuring the relative positions of the stars with a microscope.

Even then, the largest parallax we measure for the nearest star (α Centauri), is less than one second of arc: the size of a quarter (about 1 inch) seen from a distance of about 3.25 miles!

In order to deal with such large distances we have defined a distance unit: if the parallax is one second of arc, the distance is said to be one *parsec,* (a coined word consisting of the first three letters of the words parallax and second). Using the $s = r\,\theta$ equation and a baseline of 1 AU we find that

$$1 \text{ parsec} = 206{,}265 \text{ AU},$$

which is equal to 3.086×10^{13} km. The parsec is a commonly used unit by astronomers since it allows us to use the simple relationship

$$d = \frac{1}{p}$$

where p = parallax in seconds of arc
d = distance in parsecs.

One parsec is also equal to 3.26 light years.

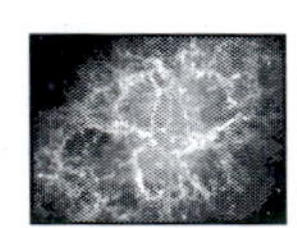

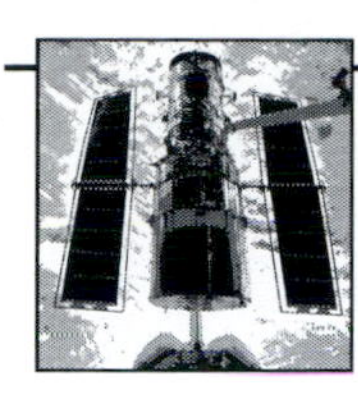

Procedure

MEASURING ANGLES

The first step is to make and calibrate your own scale for measuring angles. A transparent plastic strip marked off in 5° intervals is a simple and convenient scale, as it can be carried with you easily. The calibration procedure is described in Exercise 5 *Observing with Simple Tools.*

PARALLAX

Measuring distance using the parallax method involves measuring the angle an object appears to move through compared to a distant background when viewed from two different places. It involves applying the $s = r\,\theta$ relationship with a different pair of "knowns": now we can measure s and θ, and solve for the distance r. A common example of parallax is to hold a finger out in front of you and look at it first with one eye closed, then the other (Figure 10-3). The finger appears to move back and forth even though we know the apparent motion is really caused by looking through each eye in turn. If you measured the parallax angle θ your finger appears to move through, and the distance between your eyes s (often called the *baseline*), you could calculate the distance r to your finger.

Since

$$s = r\,\theta$$

we can solve for the distance r

$$r = \frac{s}{\theta}\ (\theta \text{ in radians})$$

or

$$r = 57\frac{s}{\theta}\ (\theta \text{ in degrees}).$$

We will use this method to measure the distance of a friend or object down a long hallway or outdoors using a distant object as a reference point.

1. Place an object or have a friend stand partway down the hall. Mark off a baseline of several feet.

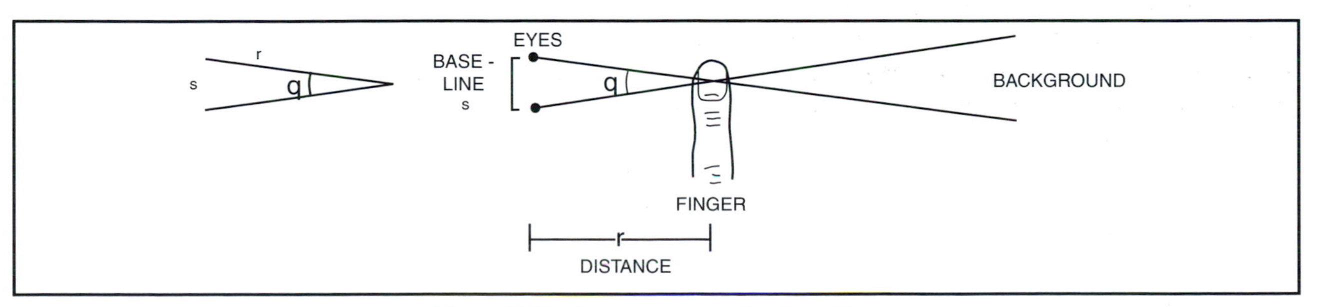

Figure 10-3. A common example of parallax.

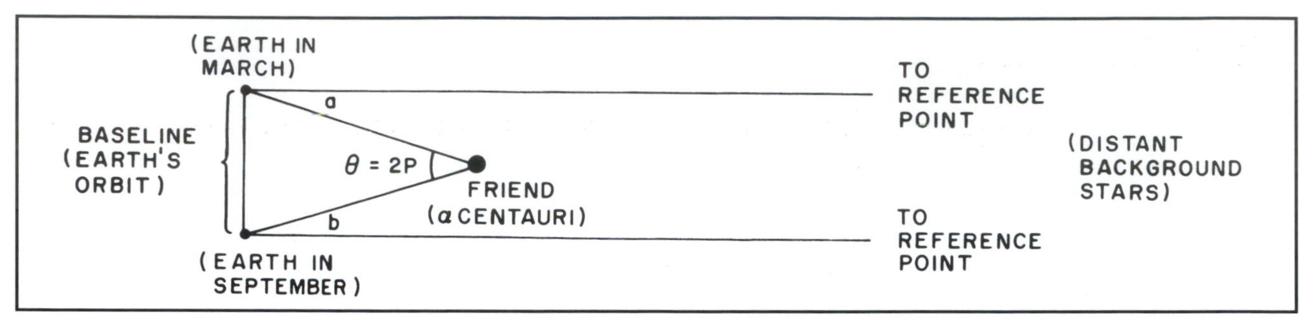

Figure 10-4. Simple parallax measurements.

2. Standing at one end of the baseline, measure the angle between your friend and some reference point as far down the hall as possible (angle a in Figure 10-4). If you can go outdoors, a distant building or radio tower will make an excellent reference object. Repeat the measurement from the other end of the baseline (angle b in Figure 10-4).
3. Add your two measured angles (a and b) to get the angle θ and find the distance to your friend using $s = r\,\theta$.

Did You Know?

In August 1989, the European Space Agency launched the Hipparcos satellite, the first satellite dedicated to measuring precise positions and motions of the stars. Operation of the satellite ceased in August 1993. The satellite observed more than 100,000 stars and was able to determine the distances to nearly 50,000 stars with a precision better than 20%. The positions of more than 20,000 stars were determined with a precision better than 10%. The published parallax accuracy, or smallest parallax angle measurable, of the satellite is 0.002 arcsec, about 50 times better than possible with the best ground-based telescopes. Improved parallax measurements led to better distance measurements for nearby stars. These measurements are the first rung of the distance-measuring ladder. Thus the accepted distances to many astronomical objects, even quite distant ones, were adjusted based on the Hipparcos results. For example, before Hipparcos the accepted distance to the Andromeda galaxy was about 2.3 million light years. Now the distance appears to be closer to 2.9 million light years. A new mission planned by the European Space Agency (Gaia), would measure the positions of more than one billion stars. As of this writing Gaia is scheduled for launch in December 2011.

Discussion Questions

1. List some possible reasons for differences in the observed and measured distances to the "model star" or friend.

2. How does the distance of the "distant" source (representing the background stars) affect parallax measurements?

3. The star Deneb is said to be at a distance of 1,600 light years. What should its parallax be? Is it probable that this figure for distance was first obtained from parallax measurements?

4. What would be the parallax of the nearest star Alpha Centauri if measurements were made from Mars? Refer to your text to obtain the distance to Alpha Centauri.

5. Calculate the maximum distance away a star may lie and still have a parallax measurable by Hipparcos.

6. Why is it advantageous to have a satellite make parallax measurements?

Exercise 11

Measuring Distances to Objects of Known Luminosity

Purpose and Processes

The purpose of this exercise is to determine the relationship between apparent brightness and source/observer separation and to use this relationship to become familiar with the process of determining the distances to special objects by measuring their apparent magnitudes. The processes stressed in this exercise include:

Observing
Using Numbers
Inferring
Using Logic
Controlling Variables

Introduction

Determining the distances to celestial objects and, thus, the size of the universe has been of foremost importance to astronomers since the very beginning of astronomy. Exercise 10 deals with parallax, which can be used to measure the distance to nearby stars but is ineffective for determining the distance to the vast majority of objects in the Milky Way, let alone the universe at large. If, however, we know the intrinsic luminosity of an object, we can use how bright the object appears to measure the distance to the object. A 100-Watt light bulb appears brighter than a 60-Watt bulb, but not if the 100-Watt bulb is a great deal further away than the 60-Watt bulb. If we **knew** that we were looking at a 100-Watt bulb, then we could tell how far away the bulb was just by how bright it appeared.

Astronomers call objects of known intrinsic brightness (luminosity) "standard candles." Cepheid variable stars represent one of the most important standard candles. [The luminosity of these stars varies periodically with the time between successive maximum luminosities dependent on that absolute brightness.] In the period from about 1910 to 1917, Henrietta Leavitt and Harlow Shapley deduced the period-luminosity relationship for Cepheid variables. By 1925 Edwin Hubble had found several Cepheid variable stars in a galaxy outside the Milky Way. Using the known period-luminosity relationship and the apparent magnitude of the stars, he was able to determine that the galaxy lay far outside the confines of the Milky Way and the universe was much larger than previously thought. Measuring distances by the use of standard candles remains one of astronomers' most important techniques.

Procedure

1. You will need a light bulb that can be covered with a box that only allows light out of a small hole, an ammeter, a dimmer switch and a photocell. Check that all these are present and that the dimmer switch changes the brightness of the bulb.
2. Separate the photocell and light bulb by about 10 cm. Adjust the box so that the light coming out of the hole falls squarely on the photocell.
3. Adjust the dimmer switch so that the bulb is on maximum brightness. Move the photocell to a position such that the ammeter reads a sizable current. In a data table record the separation between the photocell and bulb and the current.
4. Now increase the distance between the photocell and bulb in 5 cm increments, recording the separation and current for each new position. Continue increasing the separation until the current becomes undetectable or until you have eight or ten measurements.
5. Return the photocell to the initial point and dim the light so that your new current reading is half what it was at full intensity.
6. Repeat the above data taking steps, recording the data in a new table.
7. Plot both data sets on single graph of current vs. separation.

Note: This experiment can be done with photodetectors that are part of computer acquisition packages found in many physics teaching labs. See the Equipment Notes.

Optional Procedure

Astronomers often use something called a log-log plot to analyze their data. This plot is constructed by taking the logarithms of each data point and plotting these values. The beauty of the log-log plot is that it will yield a straight line if the data are related by a power law, that is $y = ax^n$, where a is any constant and n is the power in the power law relationship. For example, the volume of a cube is $V = L^3$. Here the value of n is 3 and a is 1. The volume of a sphere is $V = 4/3\pi R^3$. Here, n is once again 3 but a is $4/3\pi$. In either case, if we plotted V vs. L or R on a log-log plot we would get a straight line. The slope of the line is always n, or 3 in this case.

You can plot your data on a log-log plot to see if they are related by a power law relationship and what that relationship might be. If you use a graphing package on a computer, it should do this for you simply.

Did You Know?

Recently, astronomers have identified a certain type of supernova, Type Ia, that they believe can also be used as standard candles, with the maximum brightness of each event being nearly the same. The advantage of these supernovae is that although they are rare, they are exceedingly bright, sometimes outshining all the other stars in their host galaxy combined. Hence, they can be seen from a great distance and used to measure the distance to objects that are very far away. In the early 1990s several groups began searching for these supernovae in distant galaxies. They used these measurements to determine the expansion rate of the universe (see Exercise 14) and the how that rate was changing. If the universe began with a big bang, then the rate of expansion is expected to be slowing as the gravitational matter in the universe tries to reel all that matter back in. Just how quickly the expansion is slowing should tell us how much matter there is in the universe. Early in 1998, two separate research groups using Type Ia supernovae as standard candles reported that the expansion of the universe appears to be *speeding up*. Astronomers are now working to try to understand what this means as well as to confirm the measurements and calculations that led to this surprising result.

Portrait of Dr. Wendy Freedman

Dr. Wendy Freedman is an astronomer at the Carnegie Observatories in Pasadena, California, and a native of Toronto, Canada. During the 1980s, she and colleague Barry Madore developed techniques for measuring accurate distances to galaxies using stars known as Cepheid variables. One of her primary research interests has been aimed at measuring an accurate value for the rate at which the universe is expanding, a quantity which yields both the age and size of the universe. In the mid-1980s a panel of astronomers reviewing the top priority science for the Hubble Space Telescope designated the determination of the extragalactic distance scale and the expansion rate as one of the 'Key' (or highest priority) projects to be undertaken and completed by the telescope. Dr. Freedman was one of three principle investigators of a team of about thirty astronomers; this team was awarded the largest allocation of time on the Hubble Space Telescope for a period of five years. The project was completed in 2001, resolving a longstanding debate over the size of the universe. She has recently been appointed the Crawford H. Greenewalt Director of the Carnegie Observatories, the tenth director in the institution's one hundred year history.

Discussion Questions

In this experiment, the light bulb was used to represent a celestial object and the photocell was the observer.

1. Do the curves on your graph ever cross? Explain how you could use this technique to measure the distance to a given object.

2. How might this technique fail or give an incorrect distance? If you were measuring the distance to an object at the center of the galaxy, would you expect that to be easier or more difficult than measuring the distance to an object away from the center?

3. If you made a log-log plot, what value did you get for the slope of your line?

4. The accepted relationship is that intensity falls off as r^{-2} or $1/r^2$. Can you make a physical argument as to why this should be so?

Exercise 12

Distance to the Pleiades

Purpose and Processes

The purpose of this exercise is to plot a color-magnitude diagram for the Pleiades star cluster and to determine its distance. The processes stressed in this exercise include:

Using Numbers
Interpreting Data
Controlling Variables
Inferring
Predicting

Introduction

An open star cluster consists of a few hundred stars of common age and origin, loosely held together by mutual gravitation, and moving together through space. They usually are located in the main disc of a galaxy, in or near the spiral arms. Open clusters are therefore often called galactic clusters (not to be confused with clusters of galaxies).

Open clusters do not contain RR Lyrae stars, and only a few contain cepheid variables. In addition, distances cannot be determined from angular diameters because open clusters differ from one another in size and their angular diameters do not depend on distance alone. Distances to open clusters must therefore be determined by means different from those used for globular clusters. In practice, the most useful method for distance determination is to use the color-magnitude diagram.

Figure 12-1 shows a negative print of a photographic plate of the Pleiades, taken with a 400-mm lens and f/6.3 for 20 minutes. The star cluster's distance can be found by first plotting a color-magnitude diagram and then fitting the data from this graph to a standard Hertzsprung-Russel curve. The approximate diameter of the cluster can also be found by counting the number of stars in concentric equal-area rings.

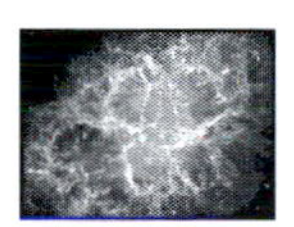

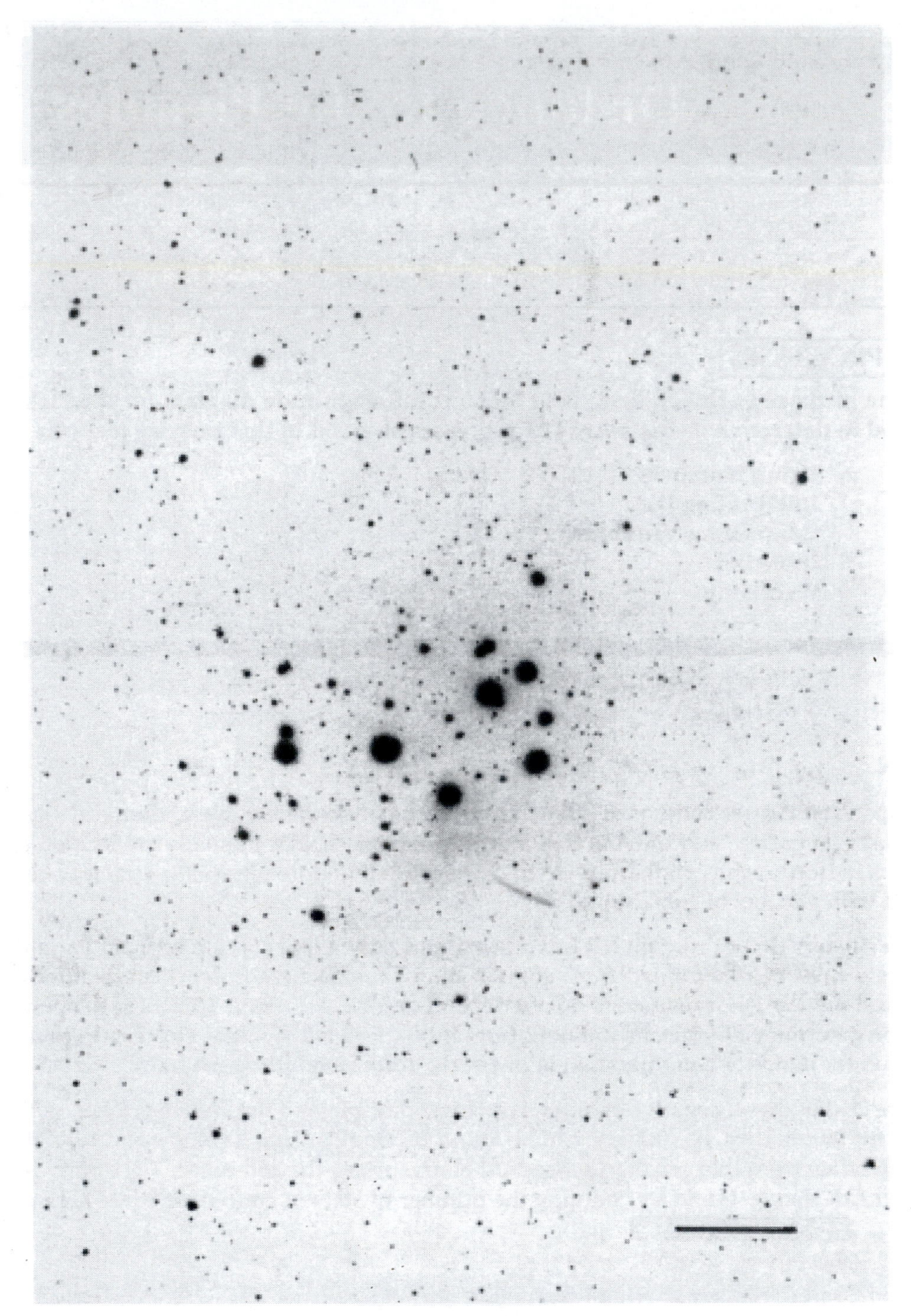

Figure 12-1. A negative print of the Pleiades. (Courtesy Dean Ketelsen, University of Iowa Observatory.)

Procedure

Table 12-1 lists colors and apparent magnitudes for many of the stars in the Pleiades cluster. The Eggen number refers to a system for identifying Pleiades stars, m_b is the blue or photographic magnitude, m_v is the visual magnitude, and $m_b - m_v$ is the color index. This information is similar to that obtained by Hertzsprung and Russell for stars of the near solar neighborhood, and also to that obtained in Exercise 26 *A Color-Magnitude Diagram of the*

Pleiades. The important difference between these data and those in the previous exercise is that here the magnitudes are given in standard magnitude units. The arbitrary magnitude scale of Exercise 26 is useful for comparing the stars within the cluster itself; however, they do NOT allow comparison with the standard main sequence to determine distance.

1. Plot a color-magnitude diagram for the Pleiades, using Table 12-1. Watch the signs on the $m_b - m_v$ values.
2. Your diagram may reveal only a portion of the main sequence. The reason for the lack of a complete main sequence in many clusters is twofold. The curve is often branched on the upper part of the main sequence as a result of the aging process of the stars in the cluster. The curve is often also truncated on the lower right for a different reason. Why?
3. For most clusters there is still enough of a main sequence on the diagram to provide the information needed to find the cluster's distance. On a second graph, USING THE SAME SCALE, plot an HR diagram for the stars given in Table 12-2. These are stars for which distances have been determined by independent means, so that the absolute magnitudes have been calculated. Note that here the absolute visual magnitude (M_V) and the $M_B - M_V$ color index are given.

Table 12-1. Colors and Magnitudes for the Pleiades[1]

Eggen Number	$m_b - m_v$	m_v	Eggen Number	$m_b - m_v$	m_v
3	+0.085	8.24	26	+0.038	7.85
5	+0.043	8.06	27	−0.215	5.74
7	+0.332	9.60	28	−0.197	6.41
8	+0.118	8.14	31	−0.242	4.17
9	+0.414	9.83	32	+0.526	10.42
10	−0.224	5.44	33	−0.073	7.34
11	−0.297	3.70	34	+0.209	8.09
13	+0.512	10.37	35	+0.620	10.20
15	+0.197	8.56	36	+0.343	9.27
16	−0.255	5.64	39	+0.527	10.51
17	−0.289	4.29	40	−0.148	6.81
18	+0.307	8.97	41	+0.149	8.37
19	+0.067	8.03	42	+0.339	9.44
20	+0.211	8.58	43	−0.140	6.98
21	+0.487	10.11	44	−0.046	7.64
22	+0.006	7.15	45	−0.133	7.25
23	+0.425	9.70	46	−0.001	7.75
24	+0.369	9.42	47	−0.098	6.80
25	−0.249	3.86	56	−0.031	6.93

[1]Adapted from H. L. Johnson and W. W. Morgan, *Ap. J.* 114, 522 (1951).

Table 12-2. Standard Main Sequence[1]

$M_B - M_V$	M_V	$M_U - M_B$	Type	Temp(eff)
−0.25	−2.10	−0.89	B2 V	20,300
−0.20	−1.10	−0.70	B3 V	18,000
−0.15	−0.30	−0.52	B5 V	15,000
−0.10	0.50	−0.33	B8 V	12,800
−0.05	1.10	−0.16	B9 V	11,800
0.00	1.50	0.00	A0 V	11,000
0.05	1.74	0.05	A2 V	9,700
0.10	2.00	0.08	A3 V	9,100
0.20	2.45	0.07	A7 V	8,100
0.30	2.95	0.02	F0 V	7,000
0.40	3.56	−0.01	F3 V	6,800
0.50	4.23	0.00	F7 V	6,300
0.60	4.79	0.08	G0 V	6,000
0.70	5.38	0.24	G8 V	5,320
0.80	5.88	0.43	K0 V	5,120
0.90	6.32	0.63	K2 V	4,760
1.00	6.78	0.87	K3 V	4,610
1.10	7.20	1.03	K4 V	4,500
1.20	7.66	1.13	K5 V	4,400
1.30	8.11	1.21	K6 V	4,000

[1]The location of the main sequence has been studied by numerous authors in order to make an assignment of colors, spectral types and absolute magnitudes. The following list is a tabulation from several authors as compiled by Evans in his book *Observations in Modern Astronomy.*
From D. S. Evans, *Observations in Modern Astronomy,* p. 95, 1968.

4. At least two different methods can be used to determine corresponding values of M_V and m_v from your data. Try using both methods and compare your results.
   (a) Read the apparent magnitude for a star of a given $m_b - m_v$ from one graph, and the corresponding absolute magnitude of a star of the same color from the second graph. Repeat for several other stars in the cluster, and calculate an average value of $m_v - M_V$, called the *distance modulus,* and often written simply $m - M$.
   (b) Place the color-magnitude graph over the standard main sequence plot, aligning the two color indices. Keeping these scales aligned, slide the top graph up and down until the color-magnitude data points best overlay the standard main sequence curve. Read corresponding values of M_V and m_v from the vertical axes of the graphs, and find the distance modulus.
5. Calculate the distance to the Pleiades using the equation

$$\log d = \frac{m - M + 5}{5}$$

where d = distance in parsecs
m = apparent magnitude
M = absolute magnitude.

OPTIONAL ACTIVITY

This optional section can be done in two ways. *Method A* is likely to produce better results but is quite time consuming. *Method B* is faster yet still produces good results.

Method A

You can calculate the diameter of a cluster in parsecs if you measure its angular diameter and know its distance, from the relationship

$$s = r\theta$$

where s = linear size (pc)
r = distance (pc)
θ = angular size (radians).

1. The scale of the photograph is indicated by the heavy line in the lower right-hand corner, with its length representing 30 minutes of arc. Determine the scale of the photograph in cm/arc min, and find the angular size of the cluster.
2. Calculate the cluster diameter in parsecs.
3. Assuming the cluster is roughly spherical, estimate its volume in cubic parsecs.
4. The density of stars in the solar neighborhood is approximately one star per 10 cubic parsecs. How does the density of stars in the cluster compare to the density of stars near us?

Method B

Using a drawing compass place the point at what appears to be the center of the cluster. Expand the compass until it appears that you have all the bright stars covered with a circle constructed out to the radial distance. Then determine the size of the cluster in cm. Proceed as in Steps 2-4 in *Method A*.

Name ______________________________ Date ______________

Exercise 12. Distance to the Pleiades

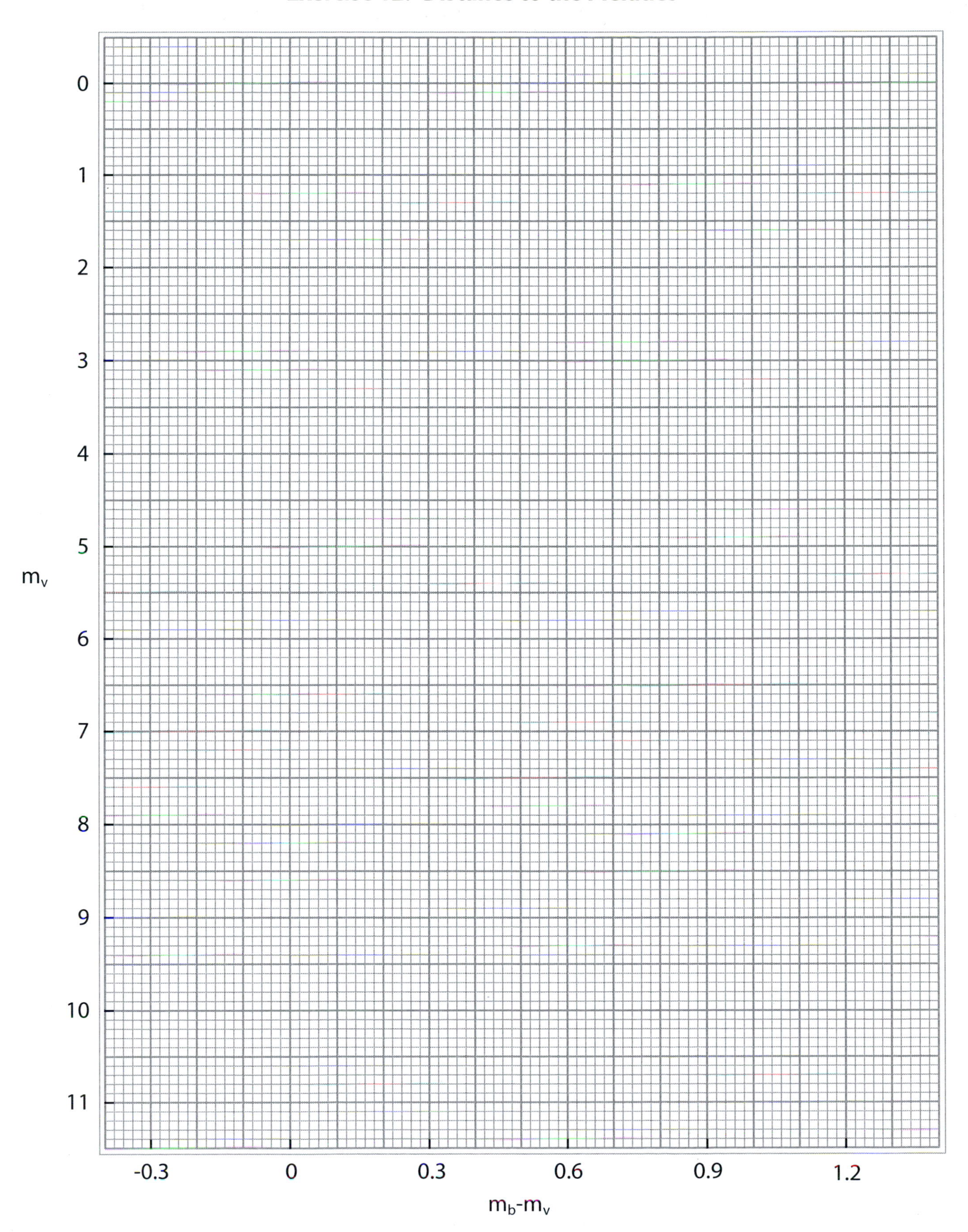

Name ______________________________ Date ______________

Exercise 12. Distance to the Pleiades

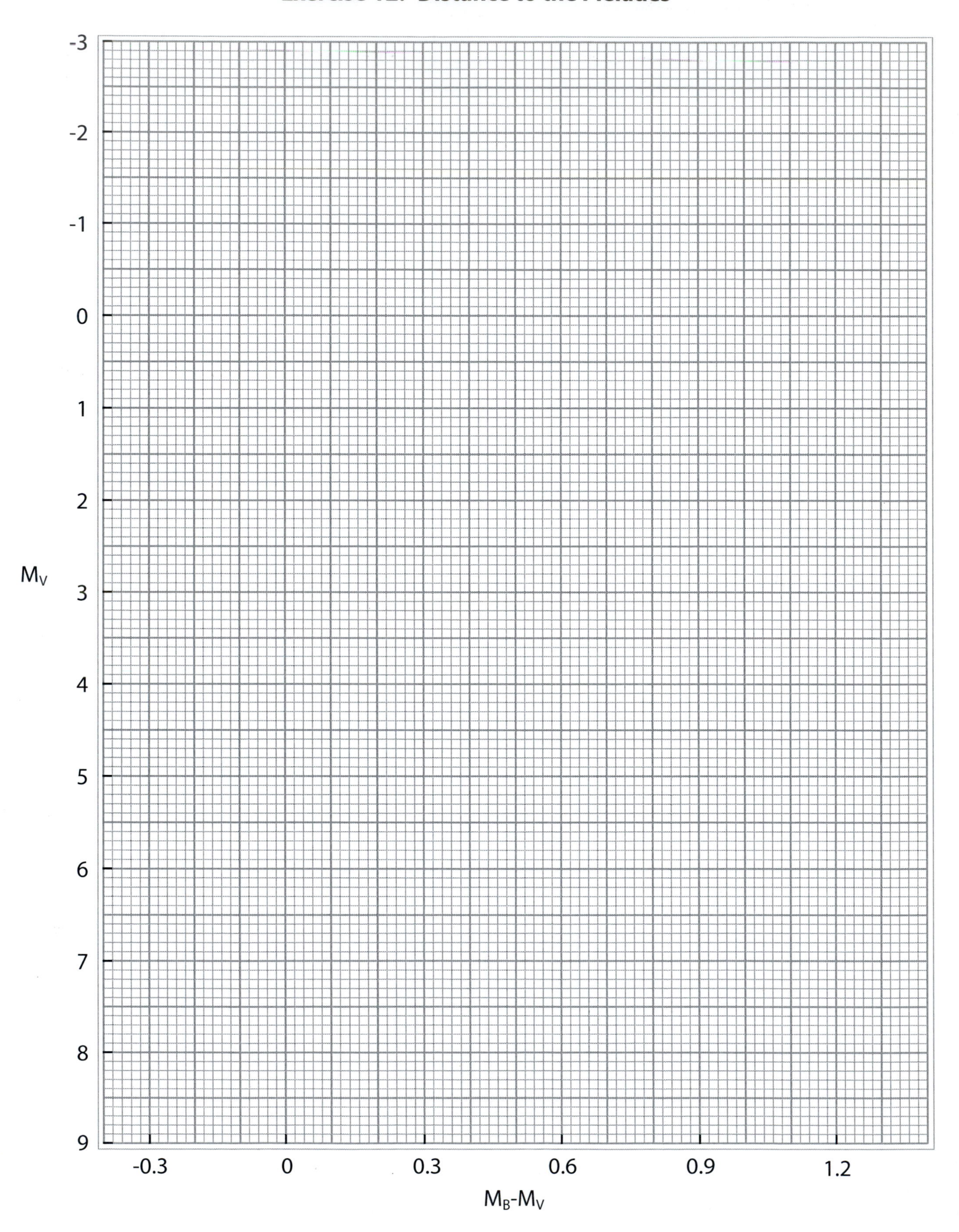

Discussion Questions

1. Which method of determining corresponding values of m_v and M_V do you think was most accurate? Estimate your uncertainty in magnitude units for each method. Would you expect the two methods to be equally valid for all clusters?

2. What assumptions are made in determining the distance to a cluster using these methods?

All the same distance away in color magnitudes

3a. What is interstellar reddening?

3.b How would interstellar reddening affect your results? Would all stars on your color-magnitude diagram be equally affected?

4. Where might this cluster be located based on its distance and your knowledge of the properties of open clusters?

Exercise

13

Distances of Cepheid Variable Stars

Purpose and Processes

The purpose of this exercise is to examine how Cepheid variable stars are used to make distance measurements in the universe.

Using Numbers
Using Logic
Inferring
Interpreting Data

References

1. Johnson, G. (2005). *Miss Leavitt's Stars.* New York: W.W. Norton and Company.
2. Leverington, D. (1996). *A History of Astronomy.* New York: Springer Publishing Company.
3. Butler, C. J. (1978). *Astronomy and Astrophysics Supplement,* 32, 83-126.
4. Percy, J.R. (2007). *Understanding Variable Stars.* London: Cambridge University Press.

Introduction

Distance measurements are among the most fundamental and difficult measurements in astronomy. Knowledge of distances to astronomical objects not only sets the scale of the universe but allows us to unlock the secret to the physical nature of objects. For example, only by measuring redshifts of quasars were we able to determine that these objects were remarkably distant and, thus, remarkably luminous. This knowledge of distance led rapidly to the model of quasars as being powered by supermassive black holes at the centers of active galaxies. See Exercises 14 and 31.

Between 1910 and 1920 two extraordinarily important distance determination techniques became available to astronomers. With the advent of the Hertzsprung-Russell(H-R) diagram and the realization that stars spend more than 80% of their lifetimes on the main sequence, it became clear that there was a nearly one-to-one relationship between spectral classification and luminosity for these main-sequence stars. Measure a star's color, determine the absolute magnitude of the star from the H-R diagram and get the distance by comparing apparent and absolute magnitudes. This technique is called main-sequence fitting or sometimes spectroscopic parallax. It is applied to open star clusters in Exercises 12 and 28.

At about the same time that it was becoming possible to measure distance using main-sequence fitting, Henrietta Leavitt was deducing that Cepheid variable stars could also be used for distance determination. She was studying Cepheid variables in the Small Magellanic Cloud (SMC) when she realized that brighter variables had longer periods. The period of the variable star is the time between successive brightness maxima or successive brightness minima. See Figure 13-1 for a sample light curve, or magnitude versus time graph, of a Cepheid variable star. Leavitt was able to discern that the SMC is quite distant (It is a satellite galaxy of our own Milky Way). As such, she could approximate that each of the variable stars was the same distance from earth, so long as the size of the SMC was relatively small compared to the distance. If any two stars lie at the same distance then the star that appears brighter to us is more luminous, so long as any scattering of the star light is the same for each. Here was the beginning of a distance-finding technique. Cepheids obey a period-luminosity relationship. If one measures the period of a Cepheid variable one can use the period-luminosity relationship to determine the absolute magnitude of the star. Comparing the absolute and apparent magnitudes yields the distance. Cepheid variable stars are valuable distance determination tools because their periods are relatively easy to measure and they are bright, allowing them to be seen at great distances. In this exercise you will plot the absolute magnitudes of Cepheid variables in the Large Magellanic Cloud (LMC), another satellite galaxy of the Milky Way, versus period. You will then use the resulting period-luminosity relationship to determine the distances to sample Cepheid variable stars.

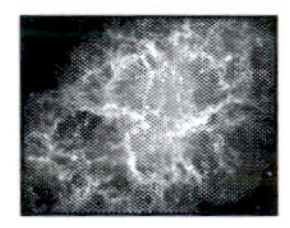

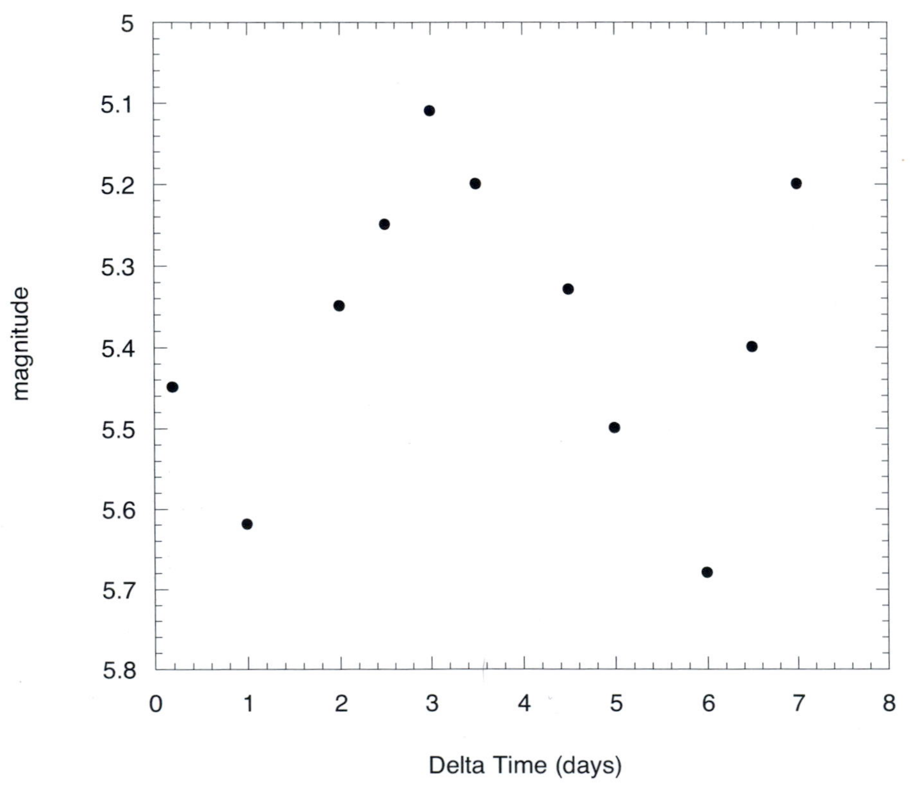

Figure 13-1. A sample light curve of Cepheid variable star with a period of about 5 days. To accurately determine the period such a variable one should measure several cycles.

Name ______________________________ Date ______________

Exercise 13. Distances of Cepheid Variable Stars

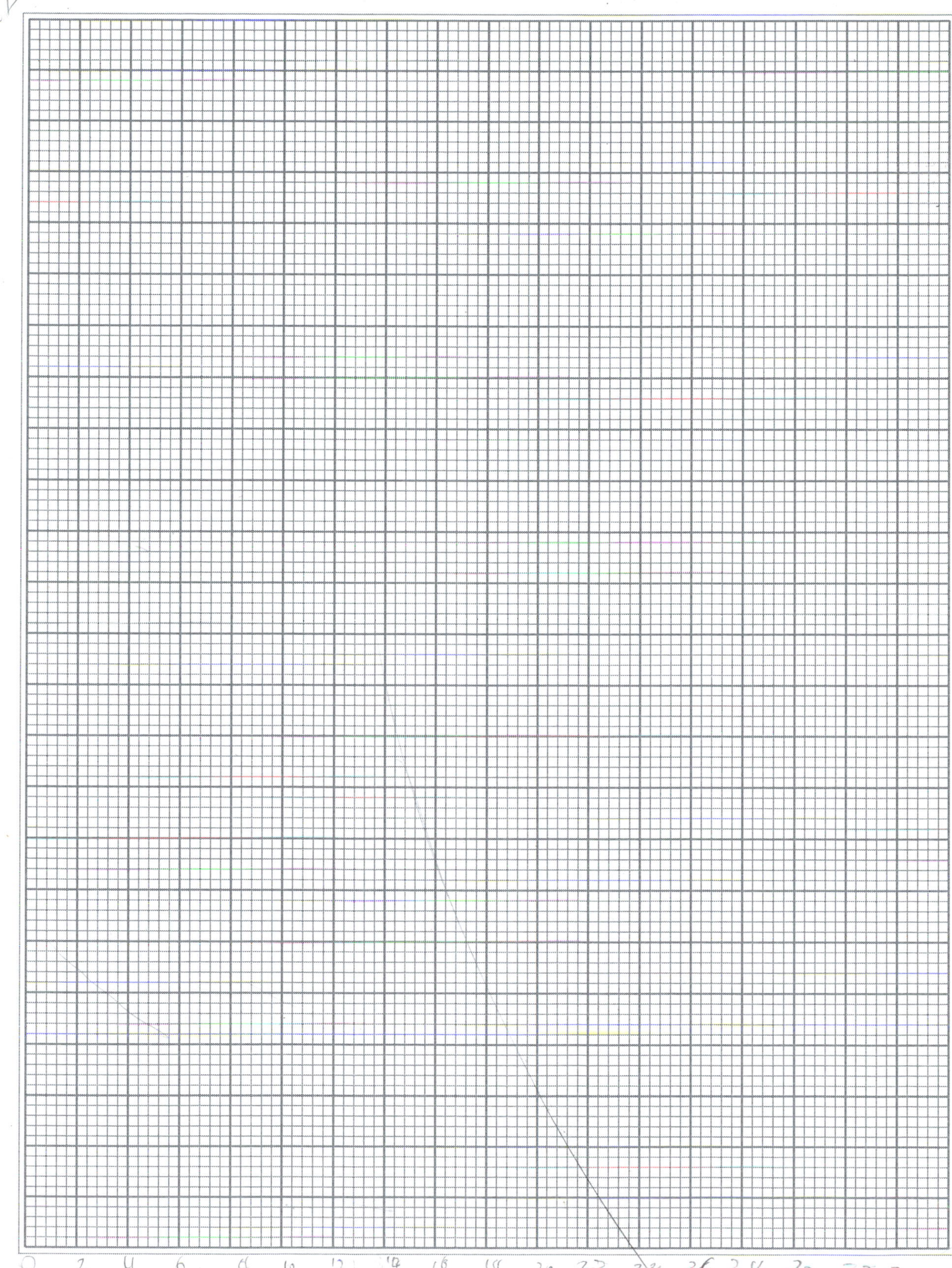

Discussion Questions

1. What possible causes might there be for any scatter found in the graph of absolute magnitude versus period that you created?

Assume data is linear
Relationship may not be linear.

2. Are the distances to Cepheids you determined more likely to be overestimates, underestimates, or are overestimation and underestimation equally likely? Explain.

Overestimated
not equally likely

3. What might be some of the advantages and disadvantages of using Cepheid variable stars to determine distance compared to using main-sequence fitting?

MSF
- cephid V'S are rare
- [illegible]
Interstellar redding

color magnitude
[illegible]
be made every [illegible]

only one graph

standard candles are designed to measure thing far away

Exercise

14

Galactic Distances and Hubble's Law

Purpose and Processes

The purpose of this exercise is to study galactic recession and Hubble's velocity-distance law. The processes stressed in this exercise include:

Designing Experiments
Using Numbers
Interpreting Data
Identifying Variables
Formulating Models
Predicting

Introduction

Determining the distances to galaxies is not simple. As a neighbor of the Milky Way, the Large Magellanic Cloud (LMC) represents an important initial step in our determination of galaxy distances but the distance to the LMC remains uncertain despite its proximity to us. A fairly large number of techniques can be used to measure the LMC distance. In recent years published estimates of the LMC distance based on several different techniques have ranged from about 42 kpc to 57.5 kpc. Improving the precision of this important measurement is of fundamental importance.

Historically, a number of methods have been used to estimate the distances to galaxies. Many of these are based on determining the apparent magnitude of a star or nebula that has an absolute magnitude that can be estimated. With a well-established period-luminosity relationship Cepheid variable stars have been an important component of this class of objects. The Hubble Space Telescope (HST) can use Cepheid variables to determine distances out to about 30 Mpc (see for example, Freedman et al., *Nature* **371,** 757, 1994). Estimates have been made using bright H II regions, brightest stars and globular clusters but all of these estimates are rather uncertain. Type Ia supernovae have become important distance indicators. With peak absolute magnitudes near −19.5 these objects can be observed out to great distances. With an apparent spread in peak magnitude values of about 0.2 these supernovae also provide fairly precise distance determinations. Recent surveys have found these objects with regularity but their transient nature means they can only be used to make distance measurements to those galaxies where they happen to appear.

Several techniques not based on the luminosities of bright objects have been developed. The Tully-Fisher relationship is an observed relationship between the between the rotational velocity of a spiral galaxy and its luminosity. This approach yields distances to spirals for which Doppler broadening of the 21-cm line can be determined. This technique can be applied out to distances of about 100 Mpc. When a galaxy is imaged with a CCD camera, pixel-to-pixel surface brightness fluctuations can be used to estimate the distance to galaxies since the size of these fluctuations is inversely proportional to the distance to a galaxy. HST might be able to determine galaxy distances to about 125 Mpc using this technique. The useful distance limit of several methods is shown in Table 14-1.

In 1929 Edwin Hubble found a correlation between the red shifts of galaxies and their distances. The apparent velocity of recession is directly proportional to the distance. Hubble used a brightest star criterion for his distance values. His law can be expressed as follow:

$$V = H_0 r$$

Where V = apparent velocity of recession in km/sec
r = galaxy distance in megaparsecs, and
H_0 = Hubble constant.

The Hubble constant is a number that sets the scale for the rate of expansion of the universe. Since this rate of expansion can change as the universe ages the Hubble constant is only constant in space and not time. The subscript on H_0 means that we are referring to the Hubble constant as it is today as opposed to, say, just after the Big Bang or perhaps five billion years ago. Hubble's **original** value for H_0 was 550 km/s/Mpc. This says that on average a galaxy recedes from us at a velocity of 550 km/sec for every million parsecs it is away from us. The Hubble constant has been re-determined several times as measurements have become more refined. Today, techniques utilizing gravitational lenses and scattering of the microwave background radiation off hot gas can provide estimates of H_0 without directly measuring the distances to galaxies.

Table 14-1. Distance Indicators for Galaxies

Object	Approximate Limiting Distance (Mpc)
Population II Red Giants	1
Globular Clusters (Average)	10
Novae (Average)	10
Novae (Brightest)	25
Brightest Blue Stars	25
Cephid variables	30
Brightest H II Regions	100
Spiral Rotation Rate	100
Surface Brightness Fluctuations	125
Type Ia Supernovae	~4,000
Brightest Cluster Galaxy	~4,000

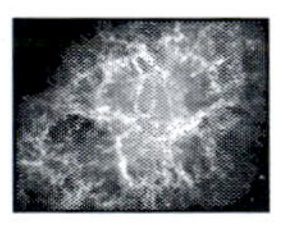

Procedure

1. Table 14-2 contains a list of clusters of galaxies and their distances. Plot these data and determine a value for the Hubble constant. Note that two independently determined distances are given for several clusters. Estimate your uncertainty in H_0 resulting from the scatter of the data.

Table 14-2. Approximate Distances and Velocities of Clusters of Galaxies

	Distances (Mpc)	Velocity (km/sec)
Pegasus I	60	4,420
	70	4,480
Perseus	83	6,360
	125	5,750
Coma	90	7,720
	185	8,350
Hercules	163	12,000
Pegasus II	225	14,850
Gemini	353	27,140
	415	26,600
Leo	310	22,300
Ursa Major	590	46,900
	698	46,400

2. Figure 14-1 shows spectra of five clusters of galaxies. Determine their red shifts and velocities of recession. A schematic of the comparison spectra identifying the various lines is given in Figure 14-2. Remember that you must:
(a) Measure the dispersion of the spectra.
(b) Measure the red shift in mm and convert to angstroms.
(c) Calculate the velocity using the Doppler equation

$$v = c\frac{\Delta\lambda}{\lambda}$$

where v = velocity
c = speed of light = 3×10^5 km/sec
$\Delta\lambda$ = shift in wavelength
λ = unshifted wavelength 3,968 Å
= Ca II, H and K 3,933 Å lines.

In Exercise 19 *Evidence of the Earth's Revolution* we used the term "dispersion." Again we will be using linear measurements (in millimeters) to tell us how much spectral lines (measured in angstroms) have been displaced from their rest position. These linear measures must be converted to angstroms. This will give you "a scaling factor" that will convert millimeters into angstroms. Identify two clear spectral lines in the spectral photographs (not in the helium comparison drawing) and measure carefully with a millimeter ruler the distance between the lines. Then locate the equivalent lines in the drawing and get the rest wavelengths of each of the lines. Subtract the shorter wavelength from the longer one. Divide this difference by the linear displacement in millimeters. This will be the dispersion of the photographic lines.

3. Using the formula from section 2 above, determine the distance to each of the clusters given. Estimate your probable uncertainty for each.

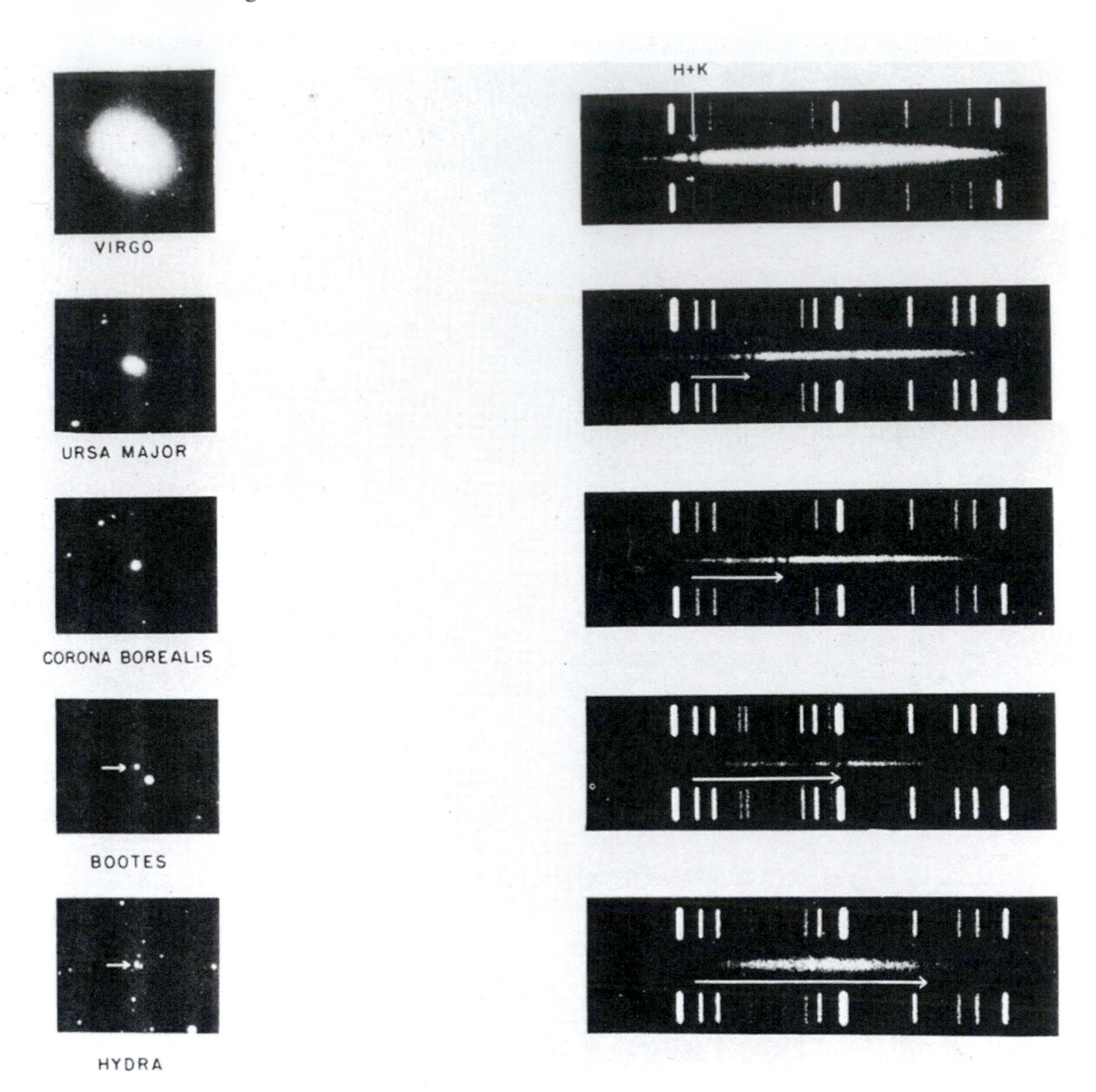

Figure 14-1. Spectra of several clusters of galaxies. (Courtesy of The Observatories of the Carnegie Institution of Washington.)

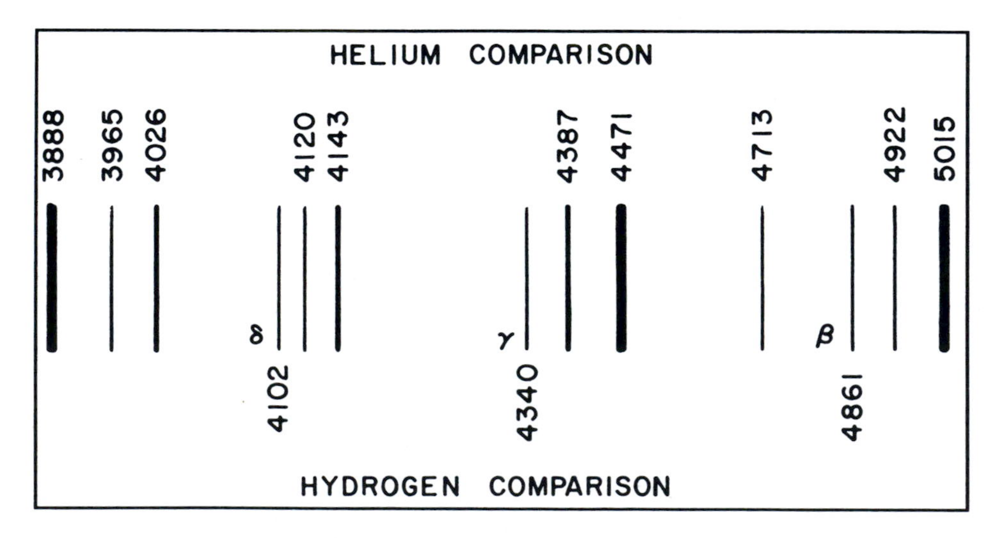

Figure 14-2. Line identification for comparison spectra.

Portrait of Anne Jump Cannon and Henrietta Swan Leavitt

At a time when women were not found among the ranks of professional scientists, Anne Jump Cannon and Henrietta Swan Leavitt were employed at the Harvard College Observatory. While Cannon worked on developing the first modern attempt to classify spectra (see Exercise 25 "Spectral Classification"), Leavitt worked on an equally fundamental problem, that of explaining why the variable stars in the Magellanic Clouds had a variation in brightness that was related to their period of variation. She found that the brighter stars in these (now known) nearby galaxies had a longer period of variation in brightness than the dimmer ones. Harlow Shapley, director of the Harvard College Observatory, recognized the importance of this *period-luminosity* relationship and later used it to map the outlines of our own galaxy. Edwin Hubble later discovered these variable stars in the Andromeda Galaxy and was able to establish that it was a distinct system of stars lying at a great distance from the Milky Way. So, in substance, it was these two under-valued women who set the stage for developing the modern method of spectral classification and for establishing the method for distance determination of objects outside of our galaxy.

Name ______________________________ Date ______________

Exercise 14. Galactic Distances and Hubble's Law

Velocity (km/sec)

Distance (Mpc)

Discussion Questions

1. What assumptions are being made about red shift in using Hubble's Law?

2. Should the effects of special relativity be considered in your results? Explain. How will this affect estimates of the size of the universe?

3. Relatively nearby galaxies can be found with a blueshifted spectrum that indicates they are moving toward us, but past a certain distance no galaxies are seen to be approaching us. Why might this be so?

4. How would the Hubble Law change if the universe were contracting instead of expanding?

PART IV
THE SOLAR SYSTEM

Exercise

15

Duration of the Sidereal Day

Purpose and Processes

The purpose of this exercise is to determine the duration of the sidereal day from a photograph of the circumpolar region of the sky. The processes stressed in this exercise include:

Using Numbers
Designing Experiments
Using Logic
Inferring

Introduction

The duration of the *sidereal day* can be defined as the time interval between two successive meridian transits of the vernal equinox. This time is based on the earth's rotation with respect to the celestial sphere or stars rather than with respect to the sun (as for the solar time). In order to measure the duration of the sidereal day we must measure accurately the apparent motion of the stars around the sky. Since this is difficult to do for a full day, it is convenient to record the motion of the stars on photographic film for a shorter, but well-known time.

The area of the sky around the north celestial pole (near Polaris) works particularly well for this purpose. The stars in this area are called *circumpolar* because they appear to circle the pole rather than rise or set.

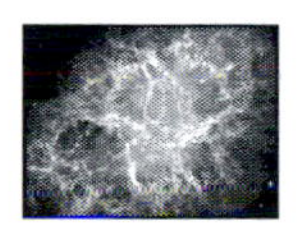

Procedure

1. If possible, take a photograph of circumpolar star trails. Attach your camera to a tripod or clamp and center the field of view near Polaris. Open the shutter for a period of 10 minutes to several hours, and record your exposure time accurately. High sensitivity film such as Tmax400 is best. Do not move your camera during the exposure.
2. If you are unable to obtain a photograph, Figure 15-1 may be used. The shutter was opened for 15 minutes, the lens cap replaced for 5 minutes, and removed again for 90 minutes to determine the direction of rotation. (These measurements were made with a stopwatch.)

Table 15-1.

Star	RA (2000)	Dec (2000)
β UMi	14 h 50.7 min	+74° 09′
γ UMi	15 h 20.7 min	+71° 50′
δ UMi	17 h 32.2 min	+86° 35′
ϵ UMi	16 h 46.0 min	+82° 02′

3. Measure the star trail arc lengths and devise a method for determining the length of the sidereal day. The brightest stars in Ursa Minor are labeled in Figure 15-1. The coordinates of these stars are given in Table 15-1. Outline your method in your lab report, and make all necessary measurements and calculations. (A sheet of polar-coordinate tracing paper might make measurements easier.)

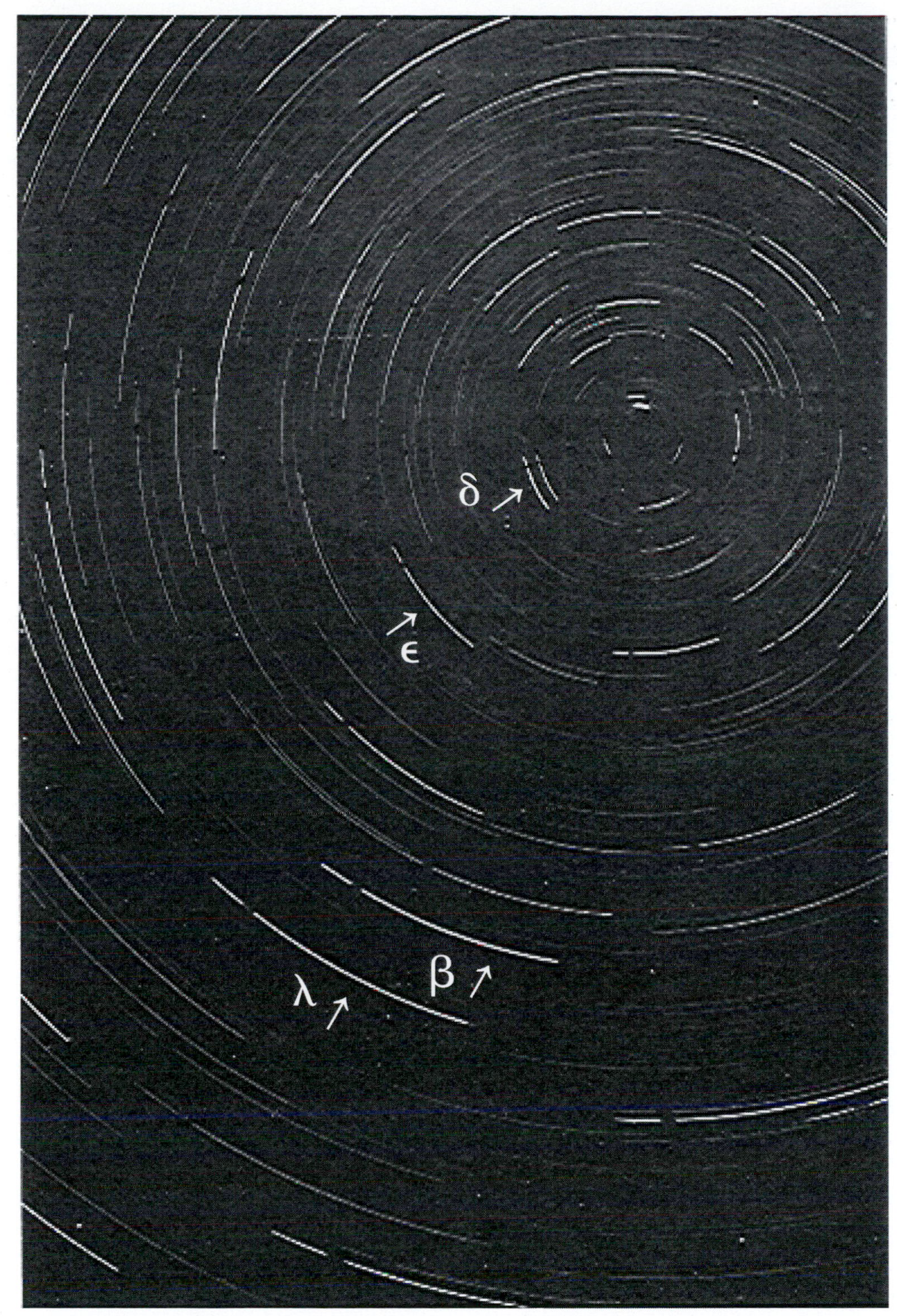

Figure 15-1. Circumpolar star trails. (Courtesy Dean Ketelsen, University of Iowa Observatory)

Discussion Questions

1. Does the measured length of the sidereal day equal the length of the solar day? Would you expect it to? Why or why not?

2. Suppose the earth's rotational period remained what we know it is, but the orbital period dropped to half what it currently is. What would happen to the lengths of the solar and sidereal days?

3. Suppose the earth rotated the opposite direction on its axis (retrograde rotation) with the same rotational period you measured in this lab. What would be the lengths of the sidereal and solar days?

Exercise

16

Lunar Features and Mountain Heights

Purpose and Processes

The purpose of this exercise is to obtain photographs of the lunar surface, study the different types of lunar features, and calculate the heights of various lunar mountains or crater walls. The processes stressed in this exercise include:

Observing
Using Numbers
Controlling Variables
Using Logic
Interpreting Data

Introduction

The moon has a great number of surface features available for observation. The large maria are visible with the naked eye and craters are easily seen with binoculars. With the greater magnification of telescopes and satellite photographs, many other types of surface features can be identified. The shadows cast by mountain peaks and crater walls are useful in mapping the three-dimensional lunar topography. The height of such features can be calculated by triangulation after the shadow length and the local altitude of the sun have been determined.

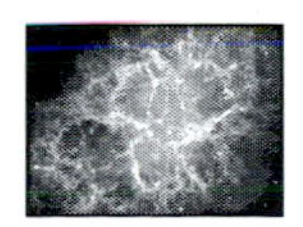

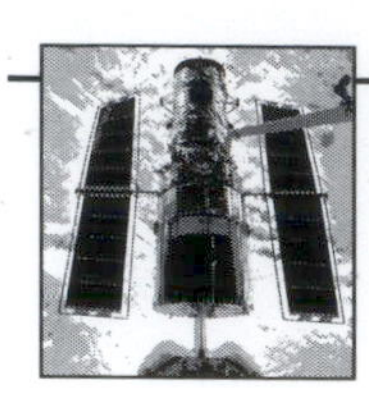

Procedure

TYPES OF SURFACE FEATURES

1. Photograph and print a series of pictures of the moon near first or third quarter, with as high a magnification as is practical. (You might want to refer to Exercise 8 *Astronomical Imaging.*) If it is not possible to take your own pictures, use the one accompanying this exercise. Lunar orbiter photographs can also be used if they are available. (See Appendix 3 *Equipment Notes.*)
2. Examine closely the photographs of the moon's surface. Sketch the large-scale features and identify by name the major maria and craters. List and describe as many different types of features as you can find, name one of each, and give its approximate coordinates.

SIZE OF LUNAR FEATURES

One can determine lunar feature sizes rather easily by measuring feature sizes on the photograph (with either a millimeter ruler or a micrometer eyepiece) and using a simple proportion.

1. Measure the diameter of your lunar image from your print.
2. Select five lunar features (maria or prominent craters) and measure their diameters to the nearest 0.1 millimeter.
3. The accepted lunar diameter is 3,476 kilometers. Using this value, set up the proportion

$$\frac{\text{MFD}}{\text{AFD}} = \frac{\text{ILD}}{\text{ALD}}$$

   where
   MFD = your measured feature diameter (mm)
   AFD = actual feature diameter (km)
   ILD = image lunar diameter (mm)
   ALD = actual lunar diameter (3,476 km).
4. Solve for the actual feature diameters in kilometers.

HEIGHT OF A MOUNTAIN OR CRATER WALL USING A LUNAR PHOTOGRAPH

Determining lunar feature heights is basically a two-step process. First the *scale height* of the feature as it appears in the scale of the photograph is found. This scale height can then be converted to the real feature's height on the moon knowing the radius of the moon and the moon's apparent radius in the photograph.

1. In Figure 16-1 let M be a surface feature whose height is to be measured. The *terminator* is the sunset or sunrise line on the surface of the moon. It is not a well-defined line as an examination of Figure 16-4 will show. However, it is necessary to estimate its location to determine lunar feature heights.

2. To more easily understand this method of measuring scale heights, visualize rotating the moon upward so that M is on the top edge as in Figure 16-2. In the enlarged Figure 16-3 we can see that triangles TOM and MAP are similar. AP and TM are parallel lines and angle APM is equal to angle TMP. Therefore, we can set up ratios of corresponding sides

$$\frac{\text{TM}}{\text{OM}} = \frac{\text{MP}}{\text{AP}}$$

where
TM represents the distance of the feature from the terminator
OM represents the radius of the moon
MP represents the scale height of the feature
AP represents the length of the feature's shadow.
Rearranging to find the feature height we have

$$\text{MP} = \text{AP}\,\frac{\text{TM}}{\text{OM}}.$$

3. Measure AP, TM and OM to the nearest 0.1 millimeter for several features on your lunar photograph. The photograph in Figure 16-4 has a lunar radius of 267 mm. Calculate the scale heights of the features you selected.
4. Measured scale heights (in mm) may be converted to actual heights of features (in km) on the moon using the following proportion

$$\frac{\text{scale height of feature (mm)}}{\text{actual height of feature (km)}} = \frac{\text{lunar radius in photograph (mm)}}{\text{actual lunar radius (km)}}.$$

The lunar radius is 1,738 km. Calculate the heights of your features in km.

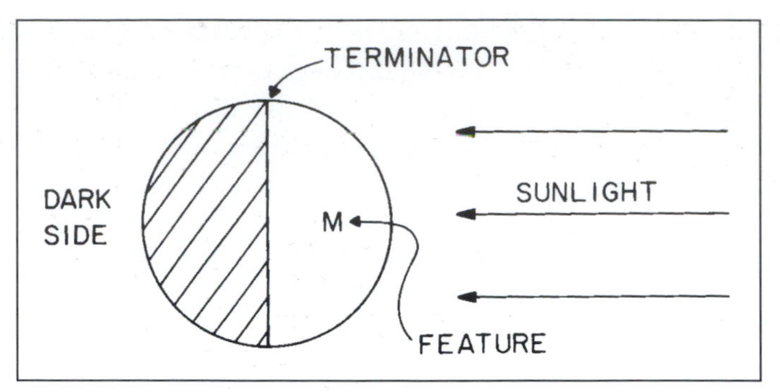

Figure 16-1. Lunar surface feature.

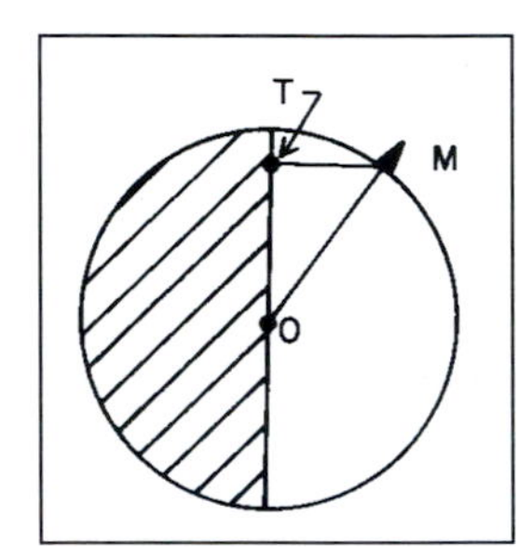

Figure 16-2. Visualized upward rotation of the moon.

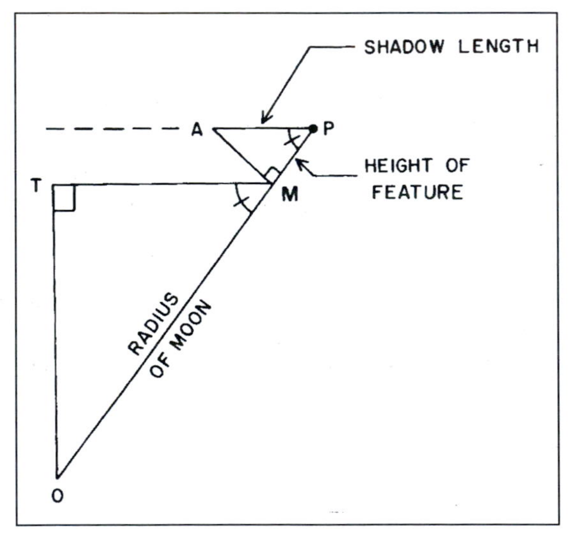

Figure 16-3. Similar triangles.

Figure 16-4. The moon near first quarter. (Courtesy Larry A. Kelsey, University of Iowa Observatory)

Exercise 17

Orbit of the Moon

Purpose and Processes

The purpose of this exercise is to become familiar with the motions of the moon from the measurement of its position relative to bright stars and planets. The processes stressed in this exercise include:

Using Numbers
Observing
Inferring
Predicting
Controlling Variables
Interpreting Data
Formulating Models

Introduction

Using simple tools to observe the position of the moon relative to a few bright stars allows an observer to track the lunar position relative to the starry background. From this information one can then deduce properties of the orbit such as how far the moon moves against the background stars in a day, how much the orbit of the moon is inclined to the ecliptic and where the moon is in its orbit when it is above or below the ecliptic. Occasionally the moon passes near a bright planet, allowing one to measure the orbital speed of the moon utilizing a series of photographs taken only minutes or a couple of hours apart. One can use the result to calculate the sidereal period of the moon.

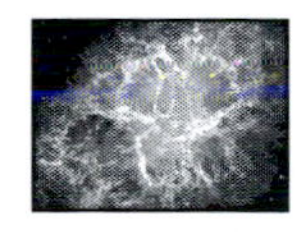

Procedure

PLOTTING THE ORBIT OF THE MOON.

1. Examine the *SC1 Constellation Chart* found in Appendix 4. Become familiar with the constellations on this chart and its system of depicting stellar magnitudes. Be certain that you understand the significance of the ecliptic, the celestial equator, right ascension, and declination.
2. Observe the moon with respect to the brighter stars as often as possible for the next four to eight weeks. Plot the position of the moon on the star chart, and record the date and time of each observation. In order to estimate angles between the moon and brighter stars a calibrated plastic strip should be used. Calibration and use of the plastic strip is described in Exercise 5 *Observing with Simple Tools*. Measure the angle between the moon and at least two (or preferably three) nearby stars. Be sure to record the angles and names of the stars being used (a log sheet for your observations is provided.) These readings may be used to triangulate the moon's position on your star chart. As an example, say we have the data given in the first two columns of Table 17-1.
   (a) Look at the declination scale on the star chart and determine the scale factor in cm per degree. Convert your degree measurements to cm.
   (b) If the moon is 20° from Betelgeuse, then it must be located somewhere on a circle of 20° radius centered on Betelgeuse, and likewise for the other stars. Using a compass, mark off an arc of the radius determined above in the general direction of the moon for each star (Figure 17-1).
   (c) The only way the moon can simultaneously be 12° from Procyon, 13° from Pollux, and 20° from Betelgeuse is for it to be at a point where the arcs for the three stars intersect. Find the point closest to where the three intersect (due to error or uncertainty in measurement they may not intersect exactly) and plot the moon there.
3. From your observations answer the following questions, supporting your answers with explanations and calculations:
   (a) In what direction does the moon travel in its orbit?
   (b) About how far does the moon move in 24 hours (expressed in degrees)?
   (c) How long does it take the moon to complete a cycle of phases? How is this time related to the orbital period of the moon around the earth?
   (d) What is the inclination of the moon's orbit relative to the ecliptic? How do you make this estimate?
   (e) If the new moon has age 0 days, estimate the age of the moon when it is at its greatest positive deviation from the ecliptic; estimate its age when it is on the ecliptic and descending; estimate its age when it has its maximum negative deviation from the ecliptic; estimate its age when it is on the ecliptic and ascending.

Table 17-1. Sample Moon Plotting Data

Star	Angle	Radius on SC1
Procyon	12°	2.0 cm
Pollux	13°	2.2 cm
Betelgeuse	20°	3.4 cm

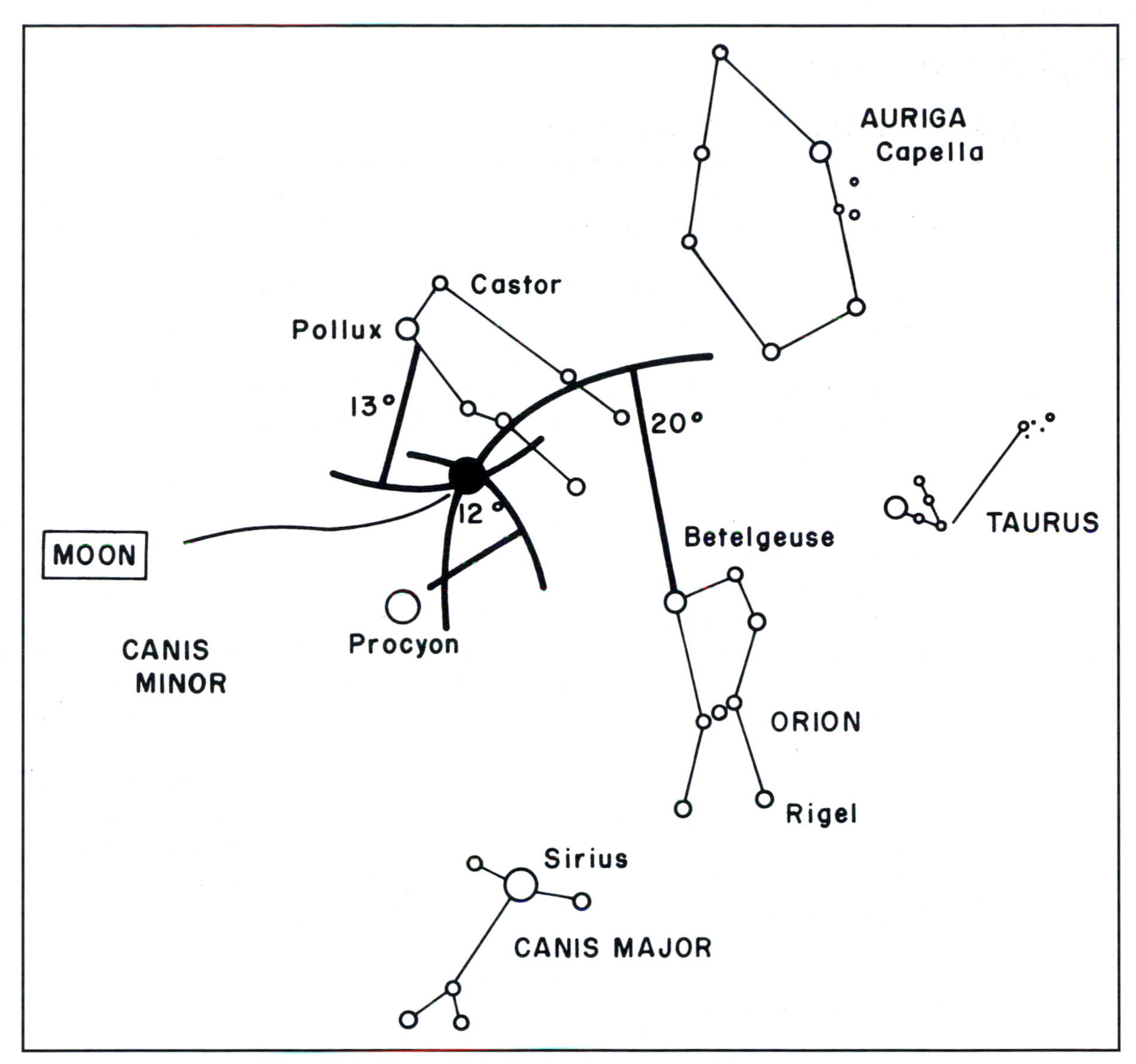

Figure 17-1. Plotting the moon's position.

(f) Use your results from part (e) to sketch orbit of the moon showing the orientation of its inclination relative to the sun and earth. That is, indicate high and low points of the orbit relative to the ecliptic and show the positions of the earth and sun.

4. Use an almanac or sky software to extend your model of the moon's orbit beyond the time you were able to observe.

(a) Repeat the estimations from part (e) in step 3 above for times approximately 3 months, 6 months, 9 months and 12 months before or after your observations.

(b) Does the phase of the moon associated with a given deviation from the ecliptic change as the earth orbits the sun? From this information would you say that the orientation of the moon's orbital tilt remain approximately fixed relative to the earth, remain approximately fixed relative to the sun or neither?

(c) Use your result 12 months before or after your observations to discuss whether you have any evidence that the orientation of the moon's orbital tilt is changing at all relative to the sun or earth.

(d) Make a sketch of one complete orbit of the earth around the sun. At four points on the orbit, sketch in the orbit of the moon around the earth. Make two of the points places where solar or lunar eclipse could occur and the other two points halfway between the eclipse points. Label the possible eclipse points.

LUNAR ORBITAL ANGULAR VELOCITY

If possible try to take your own pictures of the moon as it passes near a bright planet. The accompanying photographs (Figure 17-4) were taken of the crescent moon and the planet Venus on April 16, 1972, from Cedar Falls, Iowa. They were taken with a 135 mm lens and a single lens reflex camera on Tri-X film at f/3.5 and 1/30 second exposure time. The four pictures were taken at 7:40 P.M., 8:00 P.M., 9:00 P.M., and 10:00 P.M., respectively.

A visual examination shows the high angular velocity of the moon relative to the planet. The photos were taken facing west, and the motion of the moon relative to Venus is eastwardly. As Venus was nearly at greatest eastern elongation, it had little tangential or crosswise motion relative to the earth, and most of its motion was radial or along our line of sight (Figure 17-2). Most of the observed motion is therefore due to the motion of the moon. However, a small correction will be made later for the motion of the earth.

1. Using a sheet of tracing paper, carefully trace the outline of the moon from the first picture of Figure 17-4, and mark the position of Venus. Move the tracing paper to each successive photograph, center your moon tracing on the moon's image, and mark Venus' position for each. (You may need to rotate your drawing slightly to align the moon accurately.) Your data should produce a sketch similar to Figure 17-3. Even though it is the moon that is moving, measurements are more accurate when that motion is transferred to Venus. The image of the moon is larger, providing a clearer way to orient the sketch. Since Venus is nearly a point, it is easier to use it to trace out the straight line path of its relative motion.
2. Determine a plate scale to translate linear measurements from the photograph into angular measurements in the sky. Using a compass, trace out a full circle that best fits the circumference of the moon. The angular diameter of the moon for this date was 31 minutes of arc. Using your measured diameter of the image in millimeters, calculate a plate scale expressed in minutes of arc per millimeter.
3. Measure the distance the moon traveled in each time period on your traced diagram, and convert your values to minutes of arc. Determine the angular velocity of the moon for each time period. (Note that the motion of the moon from point A to point B was only 20 minutes.) Average your velocity values to determine the average velocity of the moon (ω) in arc minutes per hour.

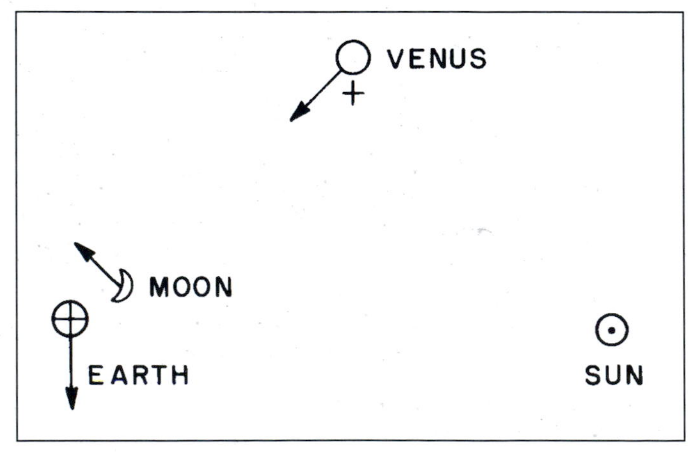

Figure 17-2. Venus at greatest eastern elongation.

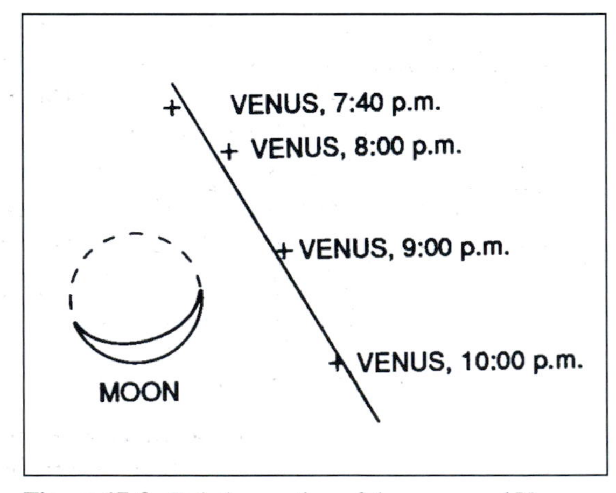

Figure 17-3. Relative motion of the moon and Venus.

Figure 17-4a. 7:40 P.M.

Figure 17-4b. 8:00 P.M.

Figure 17-4c. 9:00 P.M.

Figure 17-4d. 10:00 P.M.

Orbital motion of moon as seen against relatively stationary Venus.
(Courtesy Darrel Hoff, University of Northern Iowa.)

CORRECTION FOR EARTH'S MOTION

Refer back to Figure 17-2, which shows the relative positions of the moon, Earth and Venus (not drawn to scale). Arrows indicate the directions of motion of the earth and moon. Even though the moon is moving with the earth in its orbit, the angle formed between Earth, Venus and the moon is increasing because of the earth's orbital velocity. This angular velocity (vE) relative to Venus is roughly that of the Earth relative to the sun, which equals 360°/year or about 0.985°/day. (This value can also be expressed as 2.469/h.) Add the angular velocity of the Earth to the measured velocity of the moon (v) to give the velocity of the moon relative to a nonmoving earth (vo):

$$\omega_o = \omega + \omega_E$$

where ω_o = velocity of the moon with respect to the Earth
ω = measured velocity of the moon with respect to Venus
ω_E = velocity of Venus with respect to the Earth.

THE MOON'S SIDEREAL PERIOD

The entire orbit of the moon is 360° or 21,600′. Set up a proportion to find the sidereal period of the moon in hours and in days.

Name ______________________ Date ______________

Exercise 17. Orbit of the Moon

DATA LOG SHEET

Date	Time	Star 1	Angle	Star 2	Angle	Star 3	Angle

Discussion Questions

1. What are some of the possible sources of error in your work? Estimate their magnitudes. How much will such errors affect your value for the moon's sidereal period? Explain.

2. How would not correcting for the Earth's orbital motion in *Correction for Earth's Motion* affect your result? Calculate the orbital period of the moon WITHOUT this correction, and compare your result with the value of the synodic period of the moon. Discuss.

3. Do you expect the moon's orbital angular speed to remain constant throughout its orbit? If so, why? If not, why not and how might its orbital angular speed correlate with the apparent size of the moon in the sky?

Exercise 18

Determining the Mass of the Moon

Purpose and Processes

The purpose of this exercise is to determine the mass of the Earth's moon using data from a lunar orbiting satellite. The processes stressed in this exercise include:

Using Numbers
Interpreting Data
Using Logic
Questioning
Inferring

Introduction

Working with data obtained by Tycho Brahe, Johannes Kepler determined Three Laws of Planetary Motion:

1. The orbits of the planets are ellipses with the sun at one focus.
2. A line segment connecting the sun and a given planet sweeps out equal areas in equal time intervals.
3. $P^2 = a^3$ where P = the sidereal period of the planet in years and a = the semimajor axis of its orbit in astronomical units (A.U.). (The period describes the time of one complete revolution about the sun and the semimajor axis is defined as shown in Figure 18-1.)

Newton later generalized these laws to apply to any two bodies in orbital motion about each other and the third law reads:

$$(m + M)\,P^2 = \frac{4\pi^2}{G}\,a^3$$

where m and M are the two masses
P is the period of mutual revolution
G is the universal gravitational constant
a is the semimajor axis of their relative orbits.

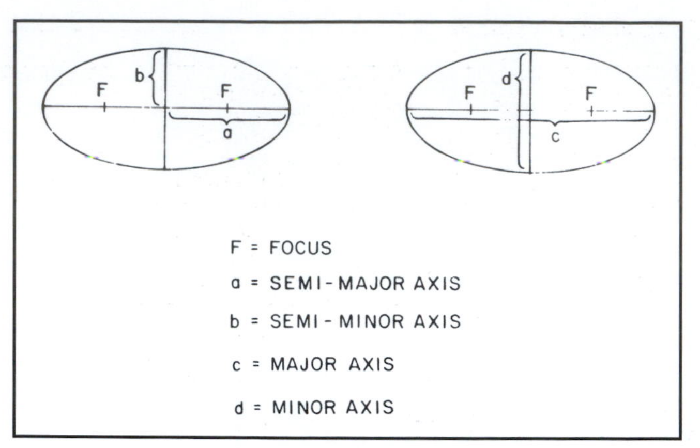

Figure 18-1. Ellipses.

This form reduces to the equation given by Kepler when we consider that a planet's mass is negligible compared to that of the sun

$$(m_{planet} + m_{sun} \approx m_{sun}):$$

$$m_{sun}P^2 = \frac{4\pi^2}{G}a^3.$$

If m is expressed in solar masses, P in years, and a in astronomical units, G will take on such a value that $4\pi^2/G$ is equal to one. So we have

$$m_{sun}P^2 = 1\,P^2 = a^3,$$

which agrees with Kepler's Law.

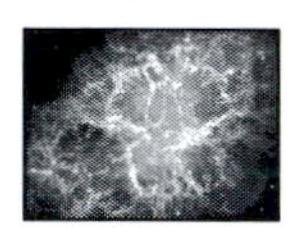

Procedure

Explorer 35 was launched from Cape Kennedy on July 19, 1967, and placed in orbit about the moon on July 22. The 230-pound NASA spacecraft carried instruments for measuring solar x-rays and energetic particles, the solar wind in interplanetary space, and the magnetic properties of the moon, as well as the interaction of the solar wind with the moon. The mission objectives were accomplished and the spacecraft continued to operate until June 1973. Table 18-1 gives a sample set of positional data of Explorer 35 in its elliptical orbit about the moon. The time interval between entries is 15 minutes, the unit of length is the radius of the moon, and the center of the coordinate system is the center of the moon.

1. Plot the data and find the major axis and foci. Verify Kepler's First Law. (Hint: Refer to the definition of an ellipse or the "string method" for drawing one. Use graph paper in the appendix.)
2. Find the semimajor axis, minor axis, semiminor axis, the eccentricity, and the period of orbit. (The eccentricity is an index of the "flatness" of the ellipse, and is defined as the distance between the foci divided by the length of the major axis.)

Table 18-1. Explorer 35 Positional Data

Elapsed Time	X (Lunar Radii)	Y (Lunar Radii)	Elapsed Time	X (Lunar Radii)	Y (Lunar Radii)
0h 00m	−3.62	1.04	6h 00m	−0.27	4.86
0h 15m	−3.46	0.63	6h 15m	−0.56	4.95
0h 30m	−3.25	0.20	6h 30m	−0.84	5.01
0h 45m	−2.97	−0.22	6h 45m	−1.12	5.03
1h 00m	−2.60	−0.65	7h 00m	−1.38	5.04
1h 15m	−2.14	−1.03	7h 15m	−1.64	5.00
1h 30m	−1.55	−1.37	7h 30m	−1.89	4.95
1h 45m	−0.85	−1.58	7h 45m	−2.14	4.87
2h 00m	−0.03	−1.59	8h 00m	−2.37	4.77
2h 15m	+0.78	−1.32	8h 15m	−2.59	4.65
2h 30m	1.45	−0.79	8h 30m	−2.80	4.50
2h 45m	1.87	−0.11	8h 45m	−2.99	4.33
3h 00m	2.09	+0.58	9h 00m	−3.17	4.14
3h 15m	2.16	1.22	9h 15m	−3.33	3.93
3h 30m	2.11	1.82	9h 30m	−3.49	3.69
3h 45m	1.99	2.35	9h 45m	−3.59	3.42
4h 00m	1.82	2.81	10h 00m	−3.69	3.15
4h 15m	1.61	3.22	10h 15m	−3.77	2.85
4h 30m	1.37	3.59	10h 30m	−3.81	2.52
4h 45m	1.11	3.90	10h 45m	−3.83	2.20
5h 00m	0.85	4.16	11h 00m	−3.81	1.83
5h 15m	0.58	4.40	11h 15m	−3.76	1.46
5h 30m	0.28	4.58	11h 30m	−3.65	1.06
5h 45m	0.00	4.74	11h 45m	−3.51	0.65

Adapted from data from James A. Van Allen, The University of Iowa. (Courtesy Goddard Space Flight Center/National Aeronautics and Space Administration)

3. Show that Kepler's Second Law is valid.
4. Using Kepler's Third Law (Newtonian form) find the mass of the moon in kilograms. Some additional information:

radius of the moon = 1,738 km

$$G = 6.668 \times 10^{-8}\ \text{cm}^3/\text{gm sec}^2$$
$$= 8.642 \times 10^{-13}\ \text{km}^3/\text{kg hr}^2$$

so that $4\pi^2/G = 4.568 \times 10^{13}$ kg hr^2/km^3.

Name ______________________________ Date ______________

Exercise 18. Determining the Mass of the Moon

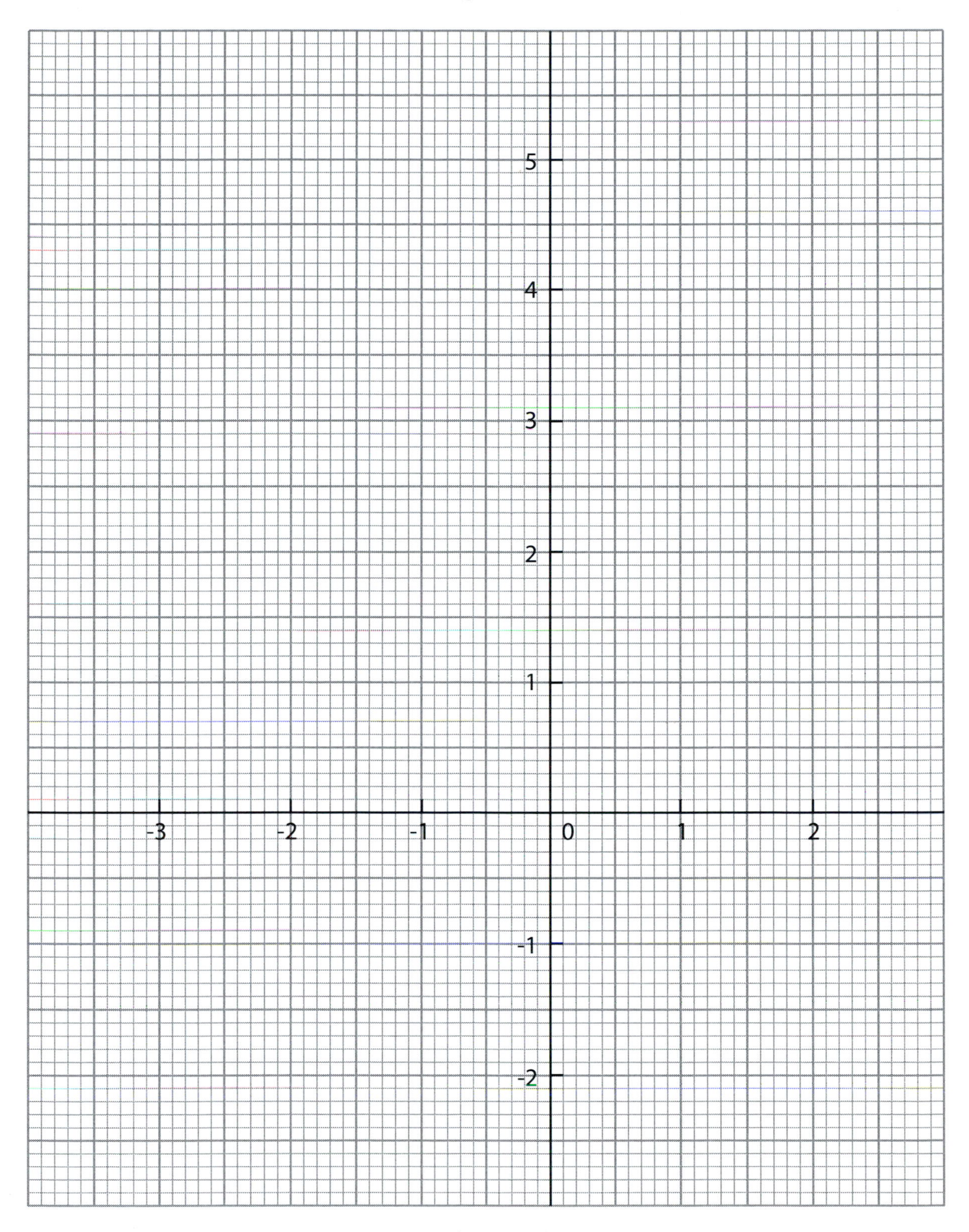

Discussion Questions

1. Sketch an ellipse with an eccentricity of 1; of 0.

2. Discuss any variations in the measurements taken to verify Kepler's First and Second Laws. What were the uncertainties in your measurements? How do they compare to the variations in your data?

3. Estimate your uncertainties in determining the mass of the moon. What has a greater effect, uncertainties in "a" or those in "P"? Why?

Exercise

19

Evidence of the Earth's Revolution

Purpose and Processes

The purpose of this exercise is to use the periodic Doppler shift of a stellar spectrum to determine the velocity of the Earth as it revolves in its orbit about the sun, and to determine the value of the astronomical unit. The processes stressed in this exercise include:

Using Numbers
Identifying Variables
Controlling Variables
Formulating Models
Questioning

Introduction

In 1842 Christian Doppler pointed out that when a light source approaches us or recedes from us, its apparent wavelength changes. His explanation assumed that light behaved as if it were made up of waves and that the wavelength of light determines the color we see. (The longer wavelengths correspond to red light and the shorter to blue or violet light.)

If a light source approaches, the waves appear to be crowded closer together to give a shorter wavelength or a blue shift. Likewise, if the motion is away from the observer, the waves will be spread out to produce a red shift (Figure 19-1).

We can apply Doppler's principle to the light we receive from stars in order to determine the relative velocity between the star and the observer. Since the Earth in its orbit is always approaching some stars and receding from others, lines in stellar spectra will show periodic Doppler shifts. We can use this fact to determine the velocity of the Earth as it revolves about the sun. With knowledge of the Earth's orbital velocity and its period, we can calculate the value of the astronomical unit.

The spectrogram in Figure 19-2 contains four separate spectra: At the top and bottom are iron spectra taken for comparison. Figure 19-2(a) gives the spectrum of Arcturus on July 1, 1939; Figure 19-2(b) is another spectrum of Arcturus, taken January 19, 1940. The comparison spectra of iron were taken by

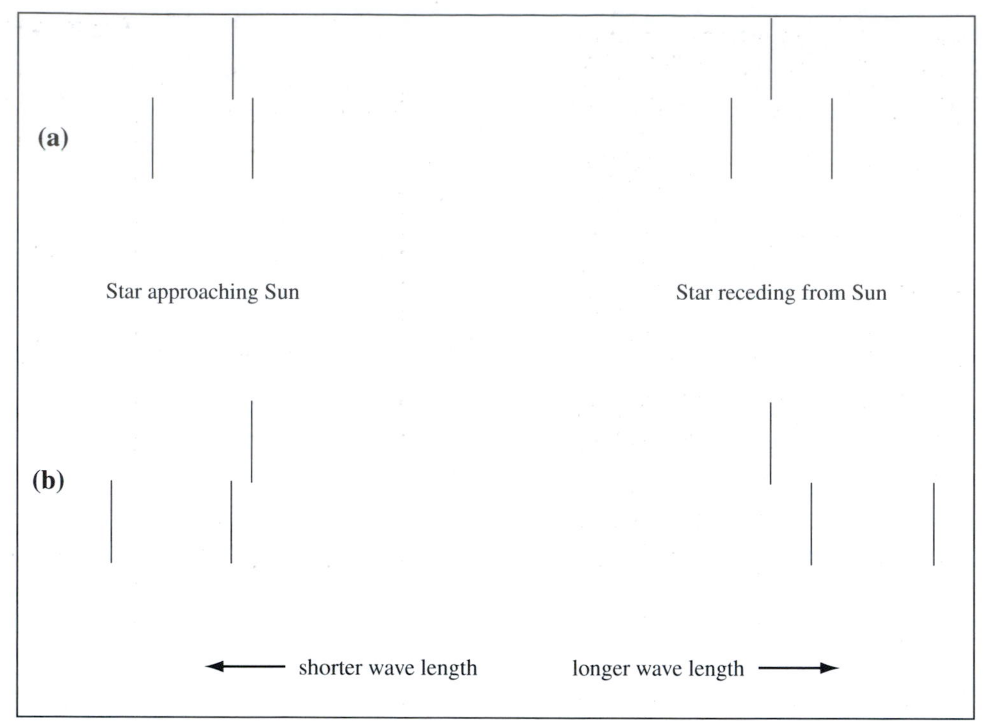

Figure 19-1. Sketches of examples of expected stellar Doppler shifts due to the combined motion of the earth and the star being observed. In case A, the speed of the star relative to the sun is smaller than the speed of the earth relative to the sun. In case B, the star's speed is larger. In either case, the speed associated with the larger shift is the sum of the star's speed and the earth's speed. The speed giving the smaller shift is the absolute magnitude of the earth's speed minus the star's speed. See Figure 19-4.

briefly illuminating an iron arc located on the telescope and are used as our rest-velocity standards. Iron was chosen because it has many easily identified lines of known wavelength. The two stellar spectra were taken six months apart at times when the Earth most rapidly approached and receded from Arcturus. We assume that the velocity of Arcturus with respect to the sun is constant.

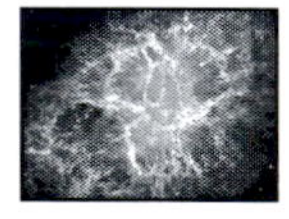

Procedure

1. A qualitative estimate of direction of motion is easily made for each spectrum by looking at several iron lines in the spectrum of Arcturus. Various iron lines have been identified for your convenience in Figure 19-2. To simplify your work, use only those lines in the spectra of Arcturus which are iron lines and identified in Figure 19-3. Note, however, that there are many other lines and several absorption bands as well.
   (a) What type of Doppler shifts do you observe in spectra (a) and (b)? Which is larger?
   (b) Since we are assuming that the star moves at a constant velocity, the apparent shifts must be because of the Earth's motion. From this information you can tell whether Arcturus is approaching or receding from the sun? Which is it doing? Does Arcturus or the earth have greater speed relative to the sun? Indicate your reasoning process. Refer to Figures 19-1 and 19-4.

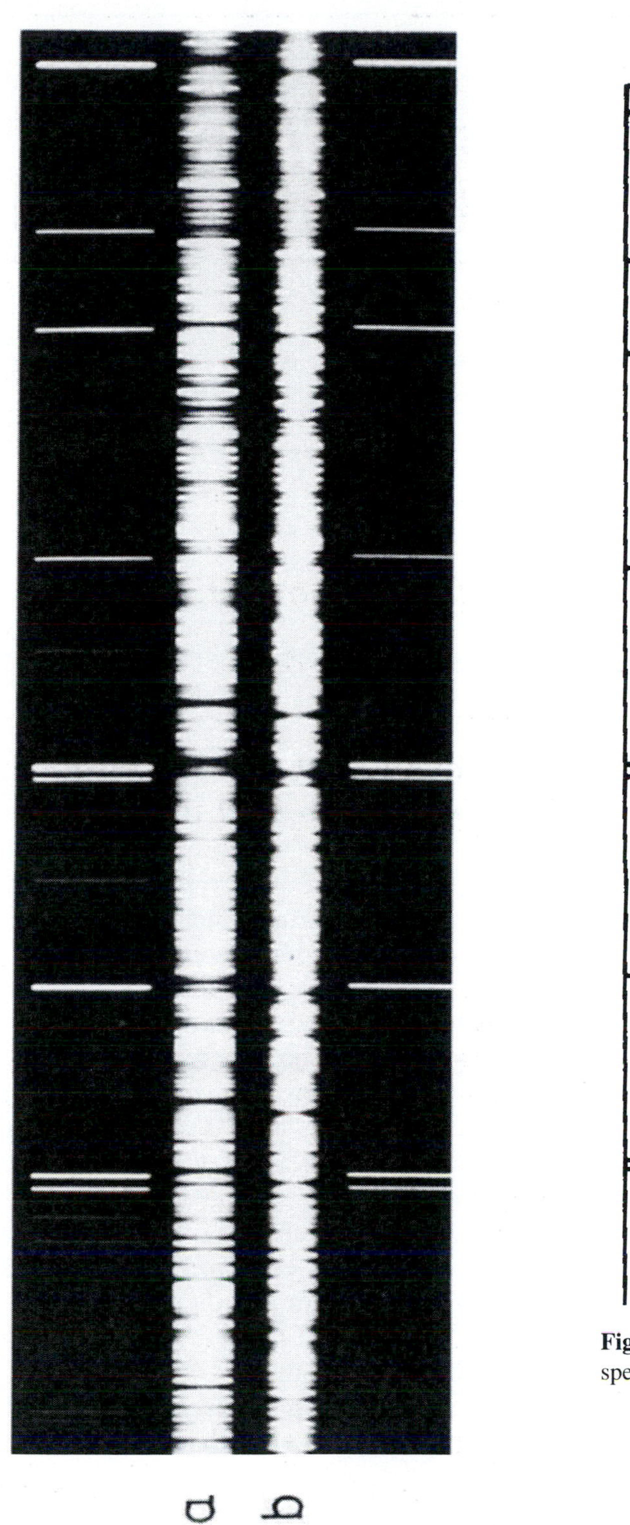

Figure 19-2. High dispersion spectra of Arcturus. (Courtesy of The Observatories of The Carnegie Institution of Washington.) Spectra (λ 4250 to λ 4310 A) of constant velocity star Arcturus taken about six months apart.
Spectra (A) 1939 July 1 and Spectra (B) 1940 January 19.

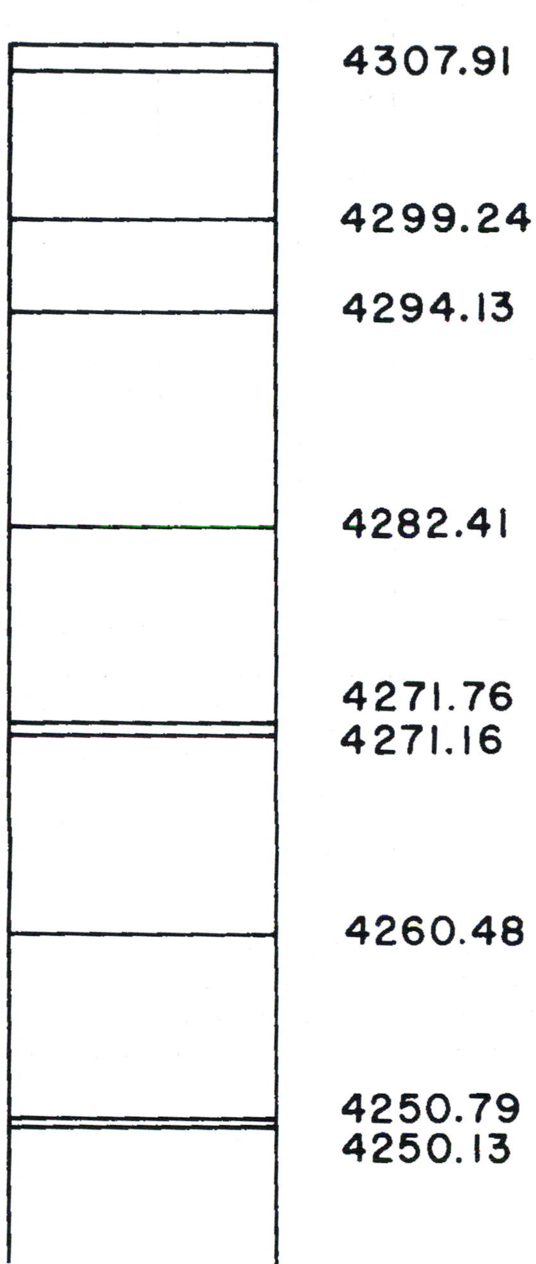

Figure 19-3. Line identification for iron arc comparison spectra. (Not to the same scale as Figure 19-2.)

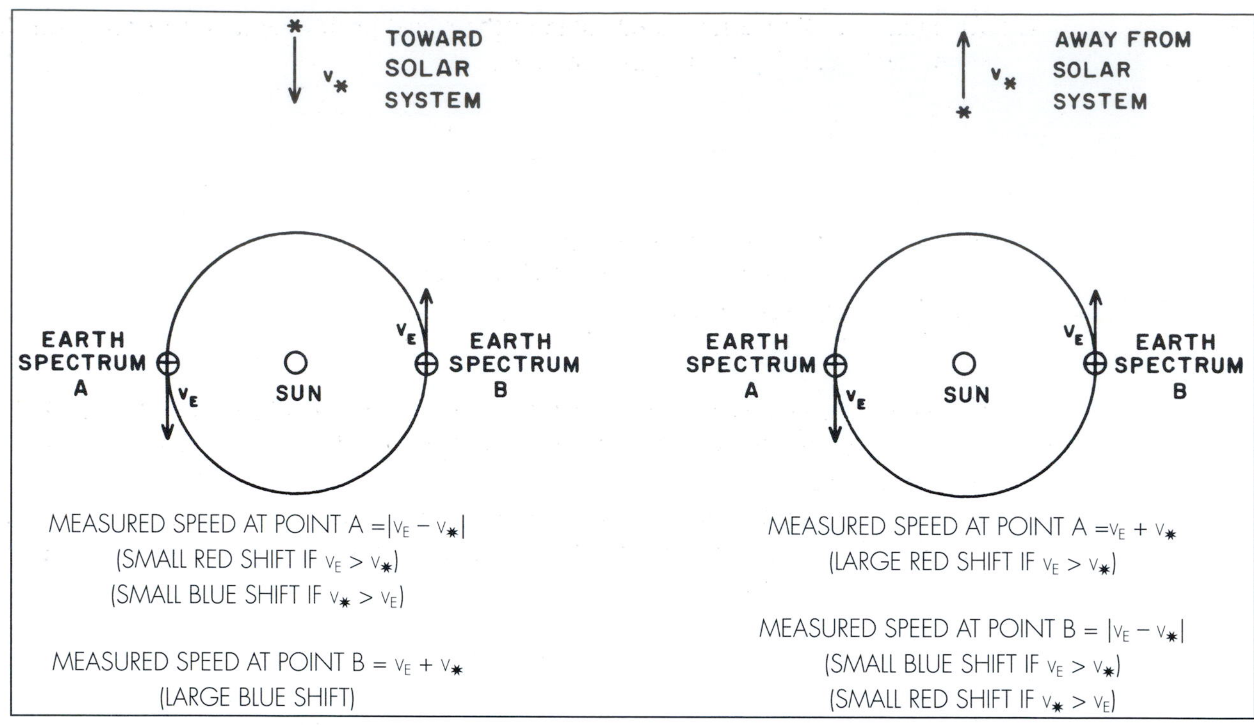

Figure 19-4. Speeds of stars approaching and receding from the solar system as measured from earth at two times six months apart.

2. (a) Measure the shifts of several prominent iron lines with respect to the corresponding iron comparison lines in each of the two spectra of Arcturus. Note that the narrow lines should be easier to measure accurately. A common method of measurement is to draw a straight, sharp line through the two iron comparison spectra and across those of Arcturus as a reference. Try to estimate your measurements to 0.1 millimeter. Measure the shift of at least 3 different lines in each spectrum.

   (b) We have measured the line shifts in millimeters directly from the spectrum but we could have made these shifts have any value we wanted simply by shrinking or enlarging the photograph. We need a scaling factor, called *dispersion,* to convert these values to angstrom units. In general, dispersion tells us how "spread out" a spectrum is, and is expressed in angstroms per millimeter.

   Determine the dispersion of these spectra (which we assume to be linear and the same for all spectra on the plate) by measuring the distance in millimeters between two known lines of the iron comparison spectrum and dividing this value into the known wavelength difference of the lines. Do this calculation for several pairs of iron lines and find an average dispersion in angstrom/mm.

   (c) Convert your millimeter measurements to angstroms, noting whether the shift is toward the red or blue.

(d) Calculate an average measured value of the speed of the star as seen from Earth for each spectrum using the Doppler equation

$$v = c\frac{\Delta\lambda}{\lambda}$$

where v = relative velocity of source and observer
c = speed of light = 3.00×10^5 km/sec
$\Delta\lambda$ = apparent shift in wavelength
λ = wavelength of comparison line (at rest).

(e) Use your measured speeds from part d and the information in Figures 19-1 and 19-4 to determine the orbital speed of the earth and the speed of Arcturus measured with respect to the sun.

(f) We can find the radius of the orbit of the Earth (the value of the astronomical unit) because we know the velocity of the Earth and the period of its orbit (one year). The relationship is expressed

$$\text{velocity} = \frac{\text{distance}}{\text{time}} = \frac{\text{circumference of orbit}}{\text{period}}.$$

By assuming a circular orbit, we have

$$v = \frac{2\,\pi\,R}{P}$$

where v = orbital velocity of the earth
R = radius of the orbit = 1 A.U.
P = period of the orbit.

Using your data, find the value of the astronomical unit assuming a circular orbit for the Earth.

Discussion Questions

1. In general, which stars does the Earth DIRECTLY approach and recede from during the course of a year? Use a star chart to name one or two.

2. Assuming its present right ascension, what would the declination of Arcturus have to be in order for these spectra to reflect maximum velocity of revolution? Explain.

3. Can the Earth's orbital velocity be determined accurately using the spectra of stars it does not directly approach and from which it does not recede? What is the necessary correction? Explain.

4. Arcturus, as seen with the naked eye, appears to be red in color. Why do we not see a change in color of the star as we change our direction from approach to recession with respect to the star?

5. How much displacement would there be in an iron spectral line of 4200 Å rest wavelength as seen by an observer on the equator of the earth. Assume that the observer takes two spectral measurements of a star on the celestial equator. One is taken in the early morning and a second after sunset. Assume the earth's rotational velocity is 4.6×10^2 m/sec and the speed of light is 3.0×10^8 m/sec.

Exercise 20

Solar Rotation

Purpose and Processes

The purpose of this exercise is to examine evidence that the sun rotates and to determine its approximate period of rotation. The processes stressed in this exercise include:

Using Numbers
Observing
Identifying Variables
Controlling Variables
Interpreting Data
Inferring
Predicting

Introduction

In the early 1600s Galileo first observed the Sun with a telescope and discovered sunspots. Note the dramatic sunspot glow in Color Image 2 in the appendix. This discovery was very disturbing to his contemporaries because the sun was thought to be a perfect "unblemished" celestial body. In his classic *Letters on Sunspots* he showed by rigorous argument that they were located on the surface of the Sun and that their motion was evidence of solar rotation, and he produced an estimate of its period of rotation. In one of the earliest quantitative determinations about the Sun, Galileo wrote, ". . . they [sunspots] have in common a general uniform motion across the face of the Sun in parallel lines. From special characteristics of this motion, one may learn that the Sun is absolutely spherical, that it rotates from west to east around its center, carries the spots along with it in parallel circles, and completes an entire revolution (sic) in about one lunar month."[1] It was not until the late 1800s, however, that an English amateur astronomer, R. C. Carrington, reported that sunspots at different solar latitudes require different times to complete one rotation.

We know now that sunspots are large areas that are relatively cooler and therefore darker than their surroundings. They are seen to last for a few days to several weeks and often are accompanied by large outbursts of optical, radio, x-ray and charged particle radiation.

Another way to determine the solar rotation rate is to use a spectrum taken with the spectrograph slit aligned along the solar equator. Since one limb is approaching us, one end of the solar lines will show a blue Doppler shift; likewise the end of the lines from the other limb will be red shifted. The part of the line

[1]From *Discoveries and Opinions of Galileo* (p. 106) by S. Drake (translator), 1957, New York: Doubleday.

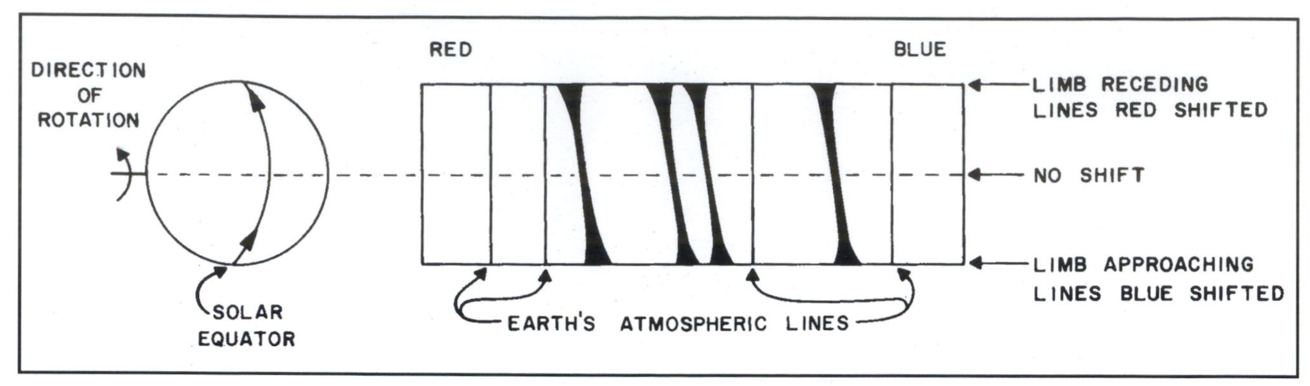

Figure 20-1. Orientation of the sun and solar spectral lines.

originating from the center of the disc will not be shifted because the rotational velocity is transverse or perpendicular to our line of sight (Figure 20-1). As a result the lines are slightly tilted with respect to the lines originating in our own atmosphere. The amount of tilt is proportional to the difference in the velocities of the two limbs, and the period can be calculated. The periods of rotation at different solar latitudes can be obtained from spectra taken at those latitudes.

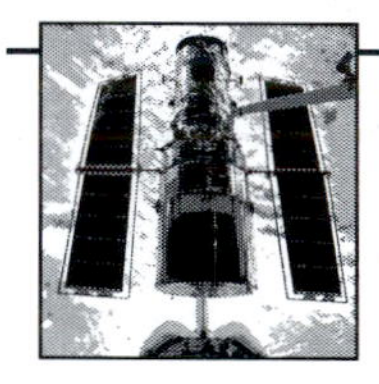

Procedure

This exercise consists of three parts: (1) examining a series of sunspot photographs to determine the direction and approximate period of solar rotation; (2) investigating solar rotation based on the examination of spectra of the approaching and receding limbs of the Sun at the equator; and (3) comparing spectra taken at several solar latitudes. Not all parts need be done; check with your instructor to determine which parts to do.

You might also want to look at Exercise 2 *Observing Exercises* for suggestions on observing the Sun and obtaining your own sunspot drawings or photographs.

SUNSPOT PHOTOGRAPHS

Figure 20-2 shows a series of photographs taken at the Hale Observatories during March and April of 1947.

1. Cover a photograph with a sheet of tracing paper and carefully trace the edge of the sun. Sketch the outlines of the prominent spots, and record the date of the picture.
2. Move your sketch to successive photographs until you find one that your spot tracing lines up with. Record the date and find the period.
3. Repeat the above procedure with at least two more sets of photographs and find an average period of rotation.
4. The photographs have the north pole of the sun at the top. Which direction is the sun rotating?

This method leads to determinating the Sun's *synodic period,* an apparent period of rotation resulting from a combination of the Sun's rotation and the Earth's orbital motion.

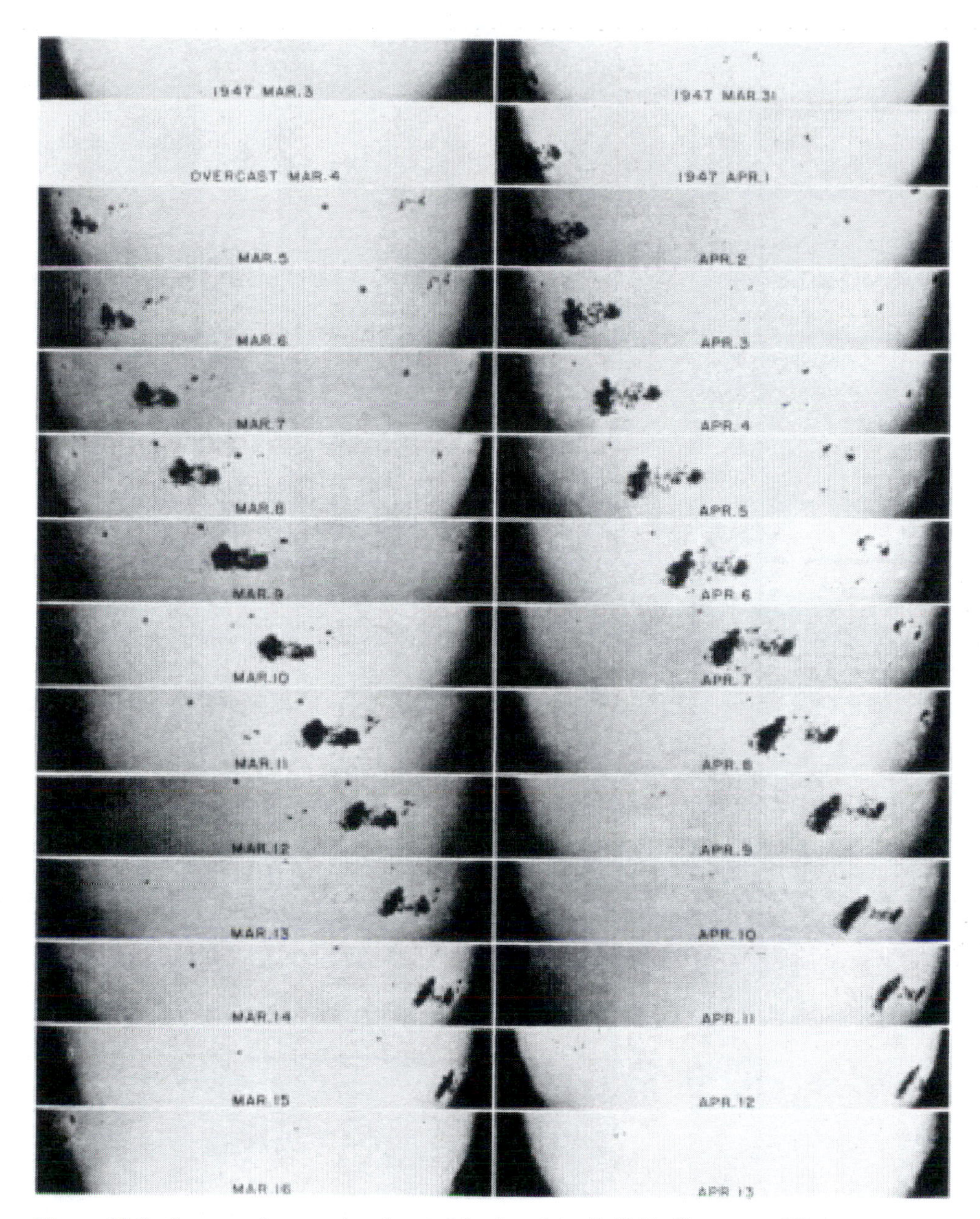

Figure 20-2. Sunspot photographs taken in March and April, 1947. (Courtesy of The Observatories of the Carnegie Institution of Washington.)

SOLAR SPECTRUM

Figure 20-3 is a photograph of a small part of the solar spectrum taken along the solar equator. You can see two types of lines: narrow lines formed by oxygen in the Earth's atmosphere; and the wide "fuzzy" solar lines. Close examination will show that the solar iron lines are slightly tilted with respect to the oxygen lines, which are not Doppler shifted at all. (You might want to review the Doppler effect and some of the details of measuring spectra in Exercise 19 *Evidence of the Earth's Revolution.*)

1. To make measurements easier, carefully draw a sharp line along and through the center of each solar line. The accuracy of your results will depend in part on being sure this line is at the center of the solar line along its entire length.

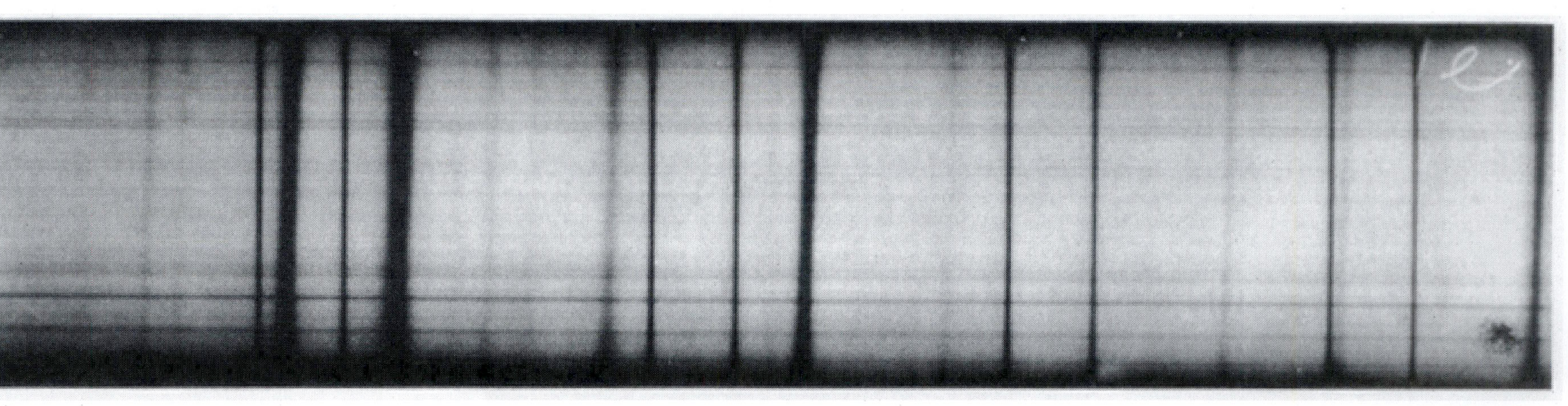

Figure 20-3. Solar spectrum at 0° solar lattitude.

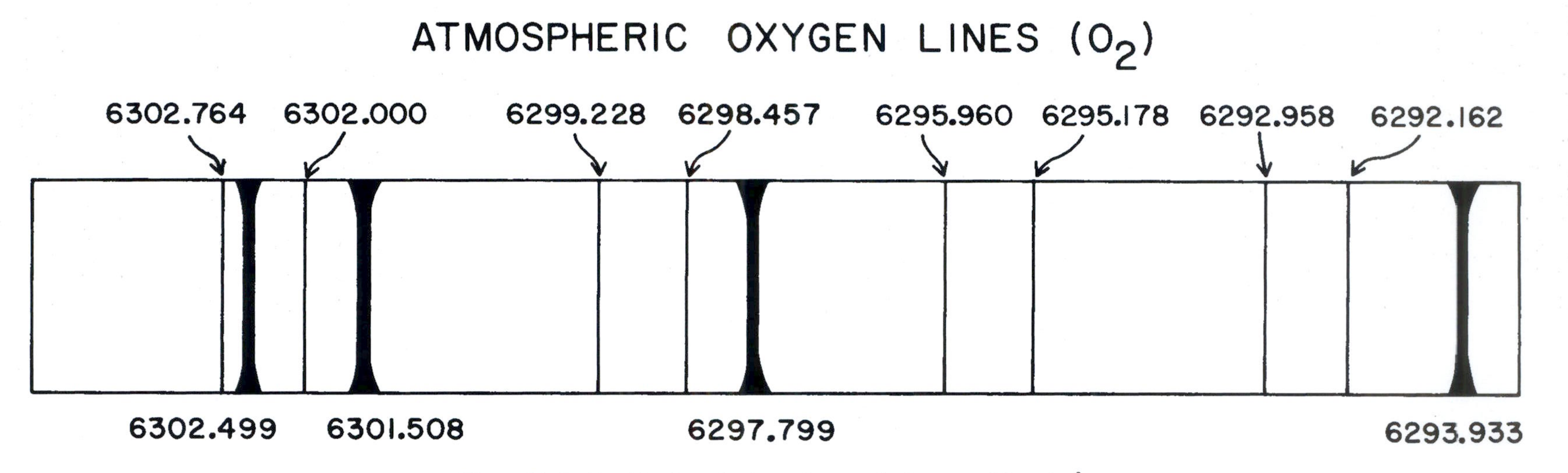

Figure 20-4. Identification of solar and atmospheric spectral lines in Å.

2. The atmospheric oxygen lines are vertical and they provide excellent references for measuring the tilt (Δx) of the solar lines: simply measure from the top of an oxygen line to the top of a solar line, and from the bottom of the same oxygen line to the bottom of the solar line. The difference is the value of Δx. Measure the tilt of several solar lines, estimating to 0.1 millimeter. Be sure to measure at the extreme top and bottom edges of the spectra. Average your values of Δx.
3. Determine the dispersion or scale of the spectrum for several sets of oxygen lines and obtain an average value. Measure the distances between the lines to 0.1 millimeter and use the wavelengths given in Figure 20-4.
4. Convert your values of Δx to Angstrom units ($\Delta\lambda$) using the average dispersion.
5. Calculate the combined velocity of the approaching and receding limbs of the Sun using the Doppler equation

$$v = c\,\frac{\Delta\lambda}{\lambda}$$

where v = the combined velocity in km/sec
c = the speed of light = 3×10^5 km/sec
$\Delta\lambda$ = Doppler shift (or tilt of lines) in Å
λ = wavelength (use a line near the center of the spectrum) in Å.

6. The rotational velocity at each limb is one-half this value because this velocity represents the difference between the two limbs. Correct for this factor of two and calculate the velocity of rotation. Remember that

$$\text{velocity} = \frac{\text{distance}}{\text{time}} = \frac{\text{circumference of Sun}}{\text{period}},$$

so

$$P = \frac{2\pi R}{v}$$

where P = the rotational period at the equator
R = the solar radius = 6.96×10^5 km
v = the measured rotational velocity.

This method determines the Sun's *sidereal period* or "true" period of rotation. Compare this value with that for the synodic rotational period from section 2 above.

ROTATION AT DIFFERENT SOLAR LATITUDES

Figures 20-5 and 20-6 were taken at solar latitudes of 60° and 90°, respectively. They were taken with the spectrograph slit aligned across the full diameter of the sun as shown in Figure 20-7.

1. Measure the rotational velocities for each latitude as you did in *Solar Spectrum,* Steps 1 through 5.
2. Since these velocities are measured at higher latitudes, the small circle at each solar latitude must be used for the circumference in calculating the period. The solar radius at this latitude (r_{60}) is found by assuming the Sun is a perfect sphere and using the relations shown in Figure 20-8. Find the solar rotational period at latitude 60°.

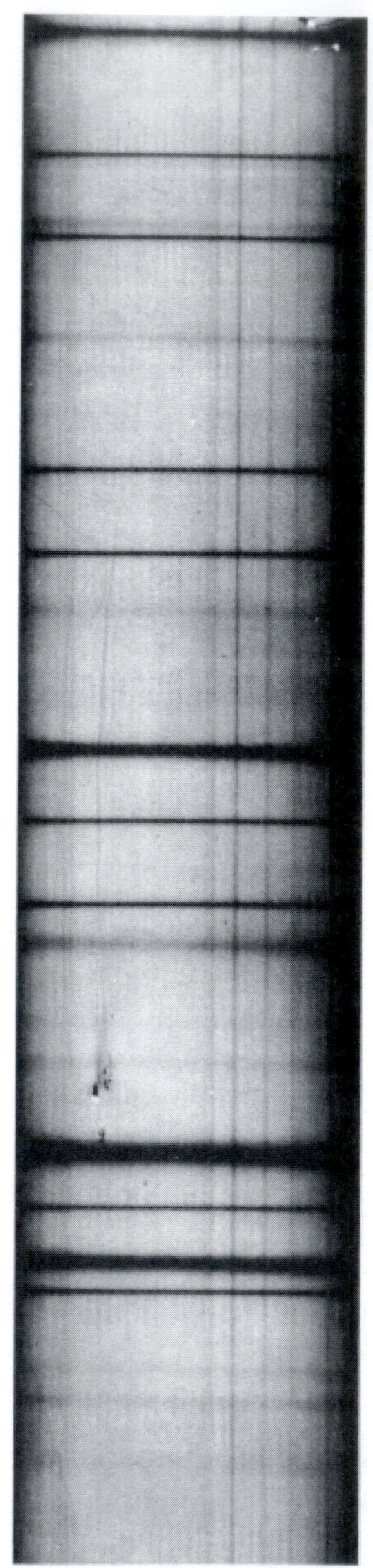

Figure 20-5. Solar spectrum at 60° solar latitude.

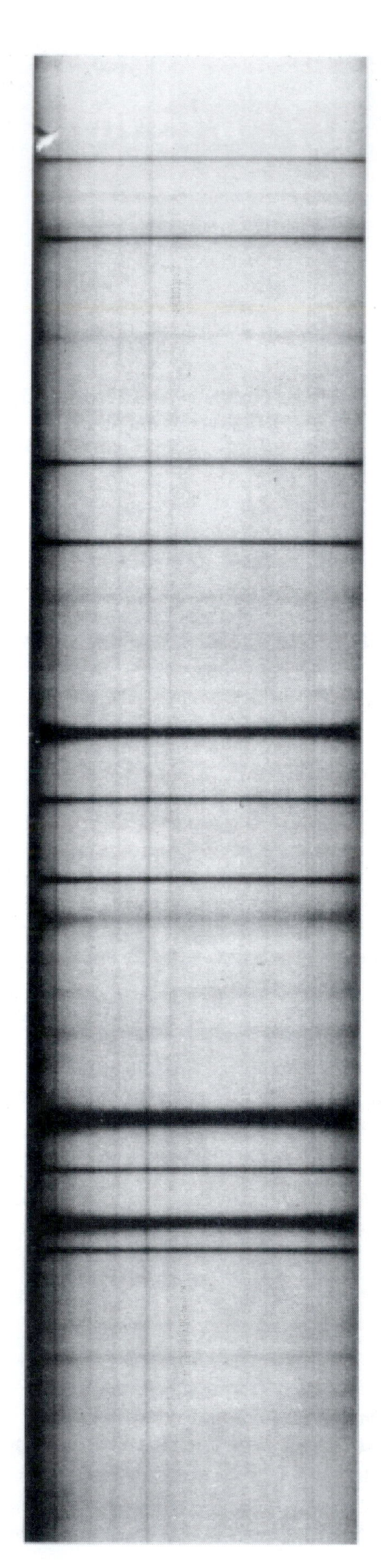

Figure 20-6. Solar spectrum at 90° solar latitude.

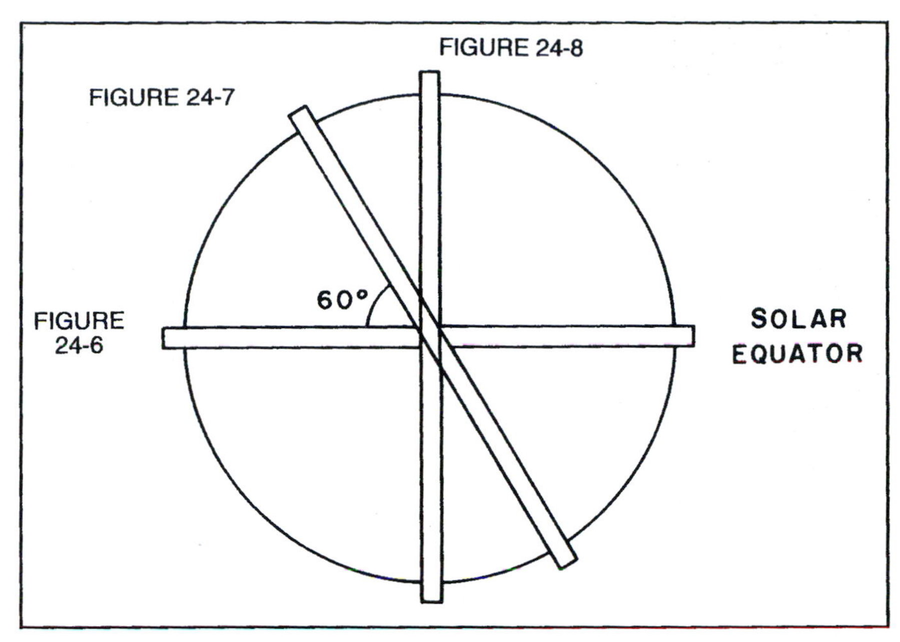

Figure 20-7. Slit alignments for solar spectra.

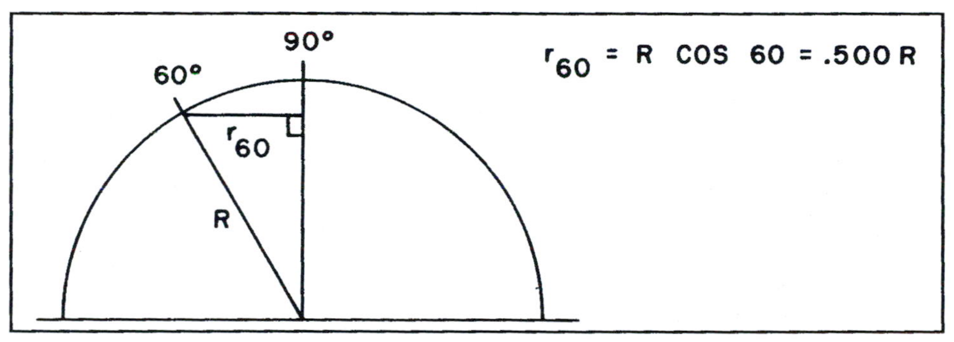

Figure 20-8. Solar circumference as a function of solar latitude.

Discussion Questions

1. What changes can you see in sunspots and in sunspot groups over a time period of several weeks?

2. Estimate and compare your errors in determining the period of the Sun's rotation using sunspots and spectra. What is the major source of error in each method?

3. The dispersion of the stellar and planetary spectra analyzed in this manual have typical values of a few angstroms or a few tens of angstroms per millimeter. Yet we find the solar dispersion is most easily expressed in units of mm/Å rather than Å/mm. Why don't we use such large dispersions for stellar spectra?

4. The solar spectral lines are much wider near the limbs than near the center. Can you explain this phenomenon?

5. Tabulate and compare class members' values for solar rotation at different latitudes. Discuss the results, including their implications and possible explanations in terms of physical models of the Sun.

Exercise 21

Measuring the Diameters of Pluto and Charon

Purpose and Processes

The purpose of this exercise is to use occultation data from the Pluto-Charon system to determine the diameter of each of the two bodies. The processes stressed in this exercise include:

Using Numbers
Controlling Variables
Interpreting Data
Formulating Models
Using Logic
Inferring

Introduction

Pluto was discovered by Clyde Tombaugh in 1930 after a year-long search of photographic plates at Lowell Observatory. He detected a small point-like object that moved against the background stars. The object was taken to be the ninth planet of the solar system and named Pluto after the ancient Greek God of the underworld. Since 1992 many similar objects have been discovered and they form the Kuiper Belt. See Exercise 22. Further study revealed that Pluto's orbit was quite eccentric and rather tilted relative to the approximate plane defined by the orbits of the other eight planets. The semi-major axis of the orbit proved to be about 40 astronomical units, yielding an orbital period of 248 years.

After the planet was discovered, attempts were made to determine its physical properties, including its size and mass. Obtaining fundamental information about the planet proved to be a difficult task. Pluto is a dim object of only about 15th magnitude. Even in the largest telescopes, Pluto does not appear as a disk. As a result, directly measuring its angular size was not possible. Through the years, many attempts to determine its size and mass led to a variety of estimates. In 1950, Gerard Kuiper used a planetary occultation of a star to estimate Pluto's diameter as approximately 6,000 km. This value appeared in the literature for a quarter century, but most astronomers understood that the value was highly uncertain.

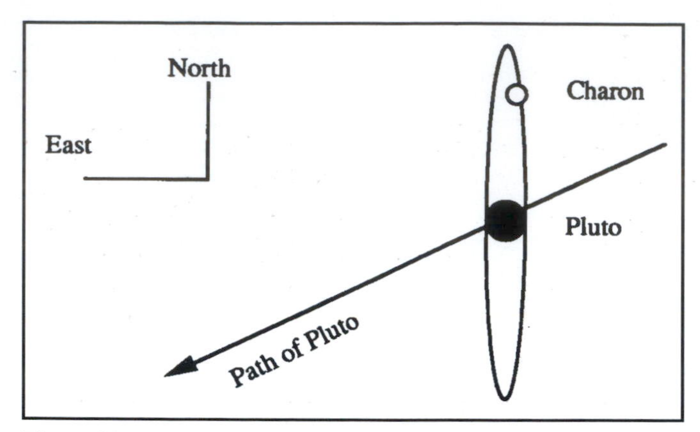

Figure 21-1. Every 124 years, the plane of charon's orbit lines up with the earth and the satellite alternately crosses in front of or behind Pluto.

Accurately determining the mass of a planet requires that the planet interact with the gravity of another object such as a nearby planet or a moon. Accurate masses for most planets are determined by their gravitational interaction with one of their moons. Pluto's mass therefore was uncertain because it did not have a detectable moon. All of this changed in 1978. While studying U.S. Naval Observatory photographic plates of Pluto, James Christy discovered a bump on the side of Pluto's images. Closer examination of the images revealed that the bump was the image of a satellite, located so close to the planet that the image of the moon and the planet blurred together. Christy named the moon Charon after the mythological boatman who carried the dead souls across the River Styx to the underworld dominated by Pluto. It was soon determined that the moon orbited the planet with a period of 6.39 days—exactly the same as the rotational period of the planet! From a knowledge of this period and the size of Charon's orbit, the mass of the planet-moon system was soon determined. (See Exercise 18 *Determining the Mass of the Moon* for an activity on determining the mass of an object from the motion of a satellite around the object.)

The problem of determining the sizes of the moon and the planet was still difficult. However, astronomers soon made a discovery about how Charon orbited Pluto that assisted in solving the size problem. Charon revolves around Pluto at almost a right angle to the path of the planet around the sun (see Figure 21-1). When Earth is in the same plane of Pluto-Charon's orbit, Charon alternately passes behind or in front of Pluto. It was soon recognized that these events could be used to accurately determine the diameters of the two bodies. This parallel geometry takes place only twice in Pluto's 248 year orbit about the Sun. By good fortune, the required geometry to observe the movement of the moon in front or in back of the planet occurred between the years 1985 and 1990.

An occultation occurs when a planet blocks the light from a more distant object. When a satellite passes in front of a planet, the event is called a transit. We can use information about an occultation of Charon by Pluto to determine the respective sizes of Pluto and Charon. Figure 21-2 shows the progress of such an occultation. First contact occurs when the limb of Charon just touches the limb of Pluto. Second contact occurs at the moment the moon completely disappears behind Pluto. The length of time between these two events is a function of the velocity of the satellite and its diameter. We can reason similarly that the time interval between first contact and when the moon begins to emerge from behind the planet (third contact) is a function of the moon's velocity and the diameter of Pluto. We can measure these time intervals and use the known velocity of Charon to measure the diameter of both Pluto and Charon.

At the time of occultation it was impossible to clearly separate Pluto and Charon in photographic images. Hubble Space Telescope was not yet in operation and adaptive optics had yet to arrive on the scene. See an HST image of the Pluto/Charon system in Color Image 10 in the appendix. Instead of directly observing the disappearance and re-emergence of Charon, astronomers used photoelectric photometers to measure the total light being received from the system. When Pluto and Charon were separated

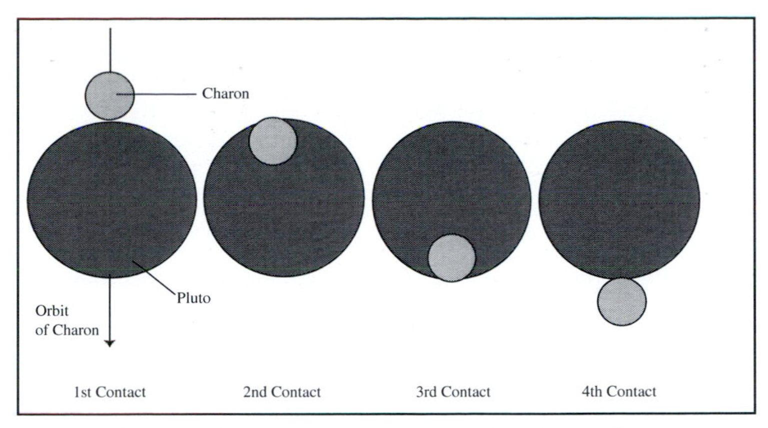

Figure 21-2. An occultation of Charon in progress. (Diagram is not to scale).

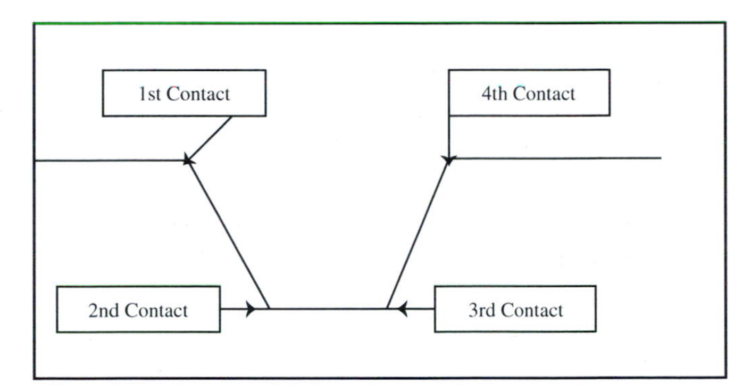

Figure 21-3. Sample light curve of Pluto occulting Charon.

from each other a photometer received light from both objects. As one or the other object disappeared, its light was blocked and the total light observed from the system decreased. Plotting this decrease in light (here called delta magnitude) as a function of time yields a light curve (magnitude versus time graph) for the occultation event. Figure 21-3 shows an ideal light curve from an occultation event. We can calculate the diameters of the two occulting objects from the elapsed times between contact points and the known relative velocities of the objects. This step requires using the formula for the distance (d) an object of speed (v) travels in time (t): $d = vt$.

In this Exercise we will use the technique just described to determine the diameters of Pluto and Charon. The technique can be used for any occulting objects, however. Only very recently has it been possible to resolve the disks of a few stars and these have been evolved giants. It is impossible to directly measure the radii of typical stars. Radii can be deduced using the Stefan-Boltzmann radiation law ($L = 4\pi R^2 \sigma T^4$) if the luminosity of the star is known. The best determinations of stellar radii come from eclipsing binary stars. Eclipsing binaries form a fairly common class of variable stars. They result from two stars in orbit around one another. The orbital planes of these systems are aligned so that we see eclipses when one star passes in front of the other, just as happens in the occultation of Pluto and Charon. The orbital speeds of the stars can be determined using the Doppler shift (see Exercise 19) and the radii of the stars can be determined fairly precisely. Visit this Exercise's web page for examples of eclipsing binary star light curves acquired with small telescopes.

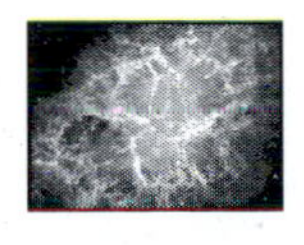

Procedure

1. Table 21-1 shows a reconstructed set of data of Pluto occulting Charon. The first column gives the time of the observation. The second column lists the change in the amount of light, expressed as *delta magnitude.* Minus values indicate a decrease in the amount of light being measured by the photometer.
2. Use the graph paper provided and plot each delta magnitude as a function of time. After completing the plot, use a straight edge to construct the light curve.

Table 21-1. Photometric Observations of Pluto Occulting Charon[1]

Time (UT)	Delta Magnitude
13:30	−0.010
13:45	−0.001
14:00	+0.012
14:15	−0.000
14:30	−0.033
14:45	−0.050
15:00	−0.097
15:15	−0.128
15:30	−0.155
15:45	−0.180
16:00	−0.218
16:15	−0.220
16:30	−0.221
16:45	−0.219
17:00	−0.220
17:15	−0.199
17:30	−0.150
17:45	−0.130
18:00	−0.092
18:15	−0.049
18:30	−0.024
18:45	−0.005
19:00	−0.006
19:15	−0.002
19:30	−0.001

[1]Reconstructed data from February 18, 1987 observations.

3. Using the data sheet provided, record your time estimate of the first, second, third and fourth contacts.
4. From the graph, determine the elapsed time between the first contact and the second contact. This is the length of time required for Charon to disappear behind the planet. This will lead to an estimate of the diameter of Charon. Record this value.
5. From your graph, determine the time from the first contact until the third contact. This is the length of time required for Charon to traverse the diameter of Pluto. Record this value on your data sheet.
6. Repeat step 4 for Charon moving from third contact to fourth contact. Record this value on your data sheet. This is a second way to get a time estimate involving Charon's diameter.
7. Repeat step 5 for Charon moving from second to fourth contact. Record this value on your data sheet. The result is a second estimate of Pluto's diameter.
8. According to recent estimates, Charon has an orbital radius (r) of 19,130 km. Assuming that the orbit of Charon is a circle, calculate the circumference (C) of Charon's orbit using the formula $C = 2\pi r$. Record this value on your data sheet.
9. It takes 6.387 days or 153.3 hours for Charon to complete one orbit around Pluto. Assume again that the orbit of Charon is a circle and calculate its orbital velocity. Record this value on your data sheet.
10. Multiply the orbital velocity of Charon by the elapsed time recorded in step 4. This is your first estimate for the diameter of Charon. Record the value on your data sheet.
11. Multiply the orbital velocity of Charon by the elapsed time recorded in step 5. This is your first estimate for the diameter of Pluto. Record this value on your answer sheet.
12. Multiply the orbital velocity of Charon by the elapsed time recorded in step 6. This is your second estimate for the diameter of Charon. Record this value on your data sheet.
13. Multiply the orbital velocity of Charon by the elapsed time recorded in step 7. This is your second estimate for the diameter of Pluto. Record this value on your data sheet.
14. Take the two values you obtained for the diameter of Charon from steps 10 and 12 respectively and average them. Record this value as your experimental diameter of Charon.
15. Take the two values you obtained for the diameter of Pluto from steps 11 and 13 respectively and average them. Record this value as your experimental diameter of Pluto.

Portrait of Clyde Tombaugh

As a young boy growing up on a Kansas farm, the late Clyde Tombaugh developed a deep interest in astronomy. Tombaugh ground his own telescope mirror and began to study the sky using an old cream separator as a mount. One of his interests was the changing appearance of the planet Jupiter, and he made detailed sketches of it from the dark skies of the prairie. After a period of time he sent these sketches to the observatory at Flagstaff. This observatory had been established at the turn of the century by Percival Lowell to study Mars. Tombaugh so badly wanted to work at such a facility that he offered to work there for no wages. He was offered a starting position but later drew a small salary for his work. He was assigned the task of locating Planet X. In the early part of the century this as yet undiscovered object was thought to be the perturbing body causing slight irregularities in the orbit of Uranus. In 1930 Tombaugh discovered the planet that was later to be named Pluto. (This photo was taken of Darrel Hoff as he visited with Tombaugh at an American Astronomical Society convention in 1986.) (Photo supplied by Darrel Hoff.)

Name ______________________________ Date ______________

Exercise 21. Measuring the Diameters of Pluto and Charon

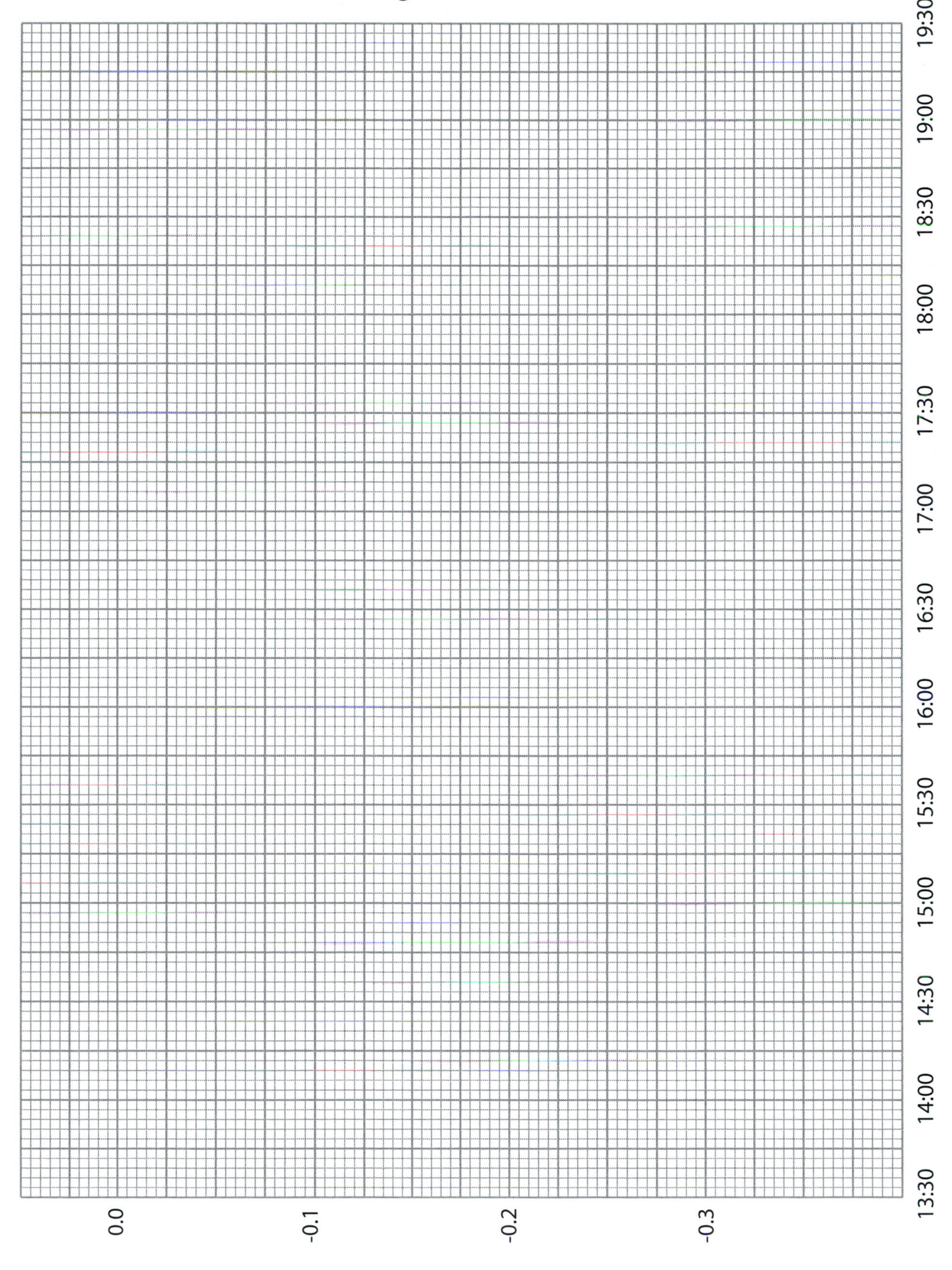

Name ______________________________ Date ______________

Exercise 21. Measuring the Diameters of Pluto and Charon

1. The times of contact occurred at:

   First ________ hours; Second ________ hours; Third ________ hours; Fourth ________ hours

2. The elapsed time between first and second contact was ______________ (Express answer in hours and decimal fractions.)

3. The elapsed time between first and third contact was ______________ (Express answer in hours and decimal fractions.)

4. The elapsed time between third and fourth contact was ______________ (Express answer in hours and decimal fractions.)

5. The elapsed time between second and fourth contact was ______________ (Express answer in hours and decimal fractions.)

6. The orbital circumference of Charon is ______________ kilometers.

7. Assuming that the orbit of Charon is a circle, its orbital velocity is ______________ km/hr.

8. My first calculated diameter of Charon is ______________ km.

9. My first calculated diameter of Pluto is ______________ km.

10. My second calculated diameter of Charon is ______________ km.

11. My second calculated diameter of Pluto is ______________ km.

12. My average value for the diameter of Charon is ______________ km.

13. My average value for the diameter of Pluto is ______________ km.

Discussion Questions

1. What are some possible sources of error in this experiment?

2. When Charon transits the planet, there is a greater decrease in the amount of light measured from the system than when Charon passes in back of the planet. What might be some reasons for this difference?

3. How might measurements made from above the Earth's atmosphere by the Hubble Space Telescope help in determining the fundamental properties of the Pluto-Charon system?

4. If Pluto had a gaseous atmosphere, how might this atmosphere change the shape of an occultation light curve and how might this change the calculations resulting from it?

5. In 2005 and 2006 astronomers used Hubble Space Telescope to discover two new moons of Pluto. Initial estimates placed these moons in the same orbital plane as Charon but 2.5 and 3 times further away, respectively. If each moon has a 120 km diameter, what would you expect for occultation depths and times between contact points when these moons occult Pluto? The orbital speed of a moon is given by $v = \sqrt{\frac{GM_{Pluto}}{r_{orbit}}}$

Exercise 22

Kuiper Belt Objects

Purpose and Processes

The purpose of this exercise is to use estimate size distribution of Kuiper Belt Objects by counting objects of various sizes observed.

Using Numbers
Interpreting Data
Inferring

Introduction

On February 18, 1930 Clyde Tombaugh discovered Pluto appearing as a 15th magnitude spot on a plate taken with the wide-field camera installed the previous year at Lowell Observatory. Three and a half weeks later the discovery was publicly announced, 149 years, to the day, after discovery of Uranus and approximately 83 and a half years after the discovery of Neptune. Had anyone imagined that the telescope would reveal a veritable plethora of planets, surely the slow pace of discovery and the meager size of Pluto cast doubt on those dreams.

Pluto was an enigma right from the start. Somewhat dense like the terrestrial planets, its average distance from the sun is 1.3 times that of Neptune, the farthest known Jovian planet. Yet, its diameter is less than half the diameter of Mercury, the smallest known terrestrial planet. In Pluto we had a planet that could not be easily classified with the others. It was theorized that, perhaps, Pluto was a moon stripped from one of the outer planets in a gravitational interaction with another planet or passing body. James Christy's discovery of Pluto's moon Charon in 1978 made that theory seem less plausible.

Our first observational evidence that Pluto might be one of many similar objects came in 1992 when Jewitt and Luu discovered a small, dense object at a distance of 42 AU from the sun. This new object, one tenth the diameter of Pluto, was the first of what have come to be known as Kuiper Belt Objects. As of 2009, more than 1000 Kuiper Belt Objects were known and more are being discovered at a rapid rate. Most known Kuiper Belt Objects range from about 30 AU to 50 AU in average distance from the sun. In 2005 the largest known Kuiper Belt Object, Eris, was discovered. It is 25 to 30% more massive than Pluto.

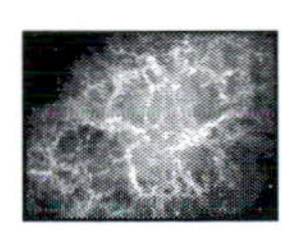

Procedure

1. Figure 22-1 shows a graph of Kuiper Belt Object radii as a function of absolute magnitude. This absolute magnitude is similar in principle to the absolute magnitude used for stars. That is, the distance to a Kuiper Belt Object is determined by its apparent motion across the sky. Its distance and apparent magnitude are then used to calculate an absolute magnitude that can be used to compare KBOs at different distances. Recall that these objects only reflect sunlight. They do not emit their own visible radiation. Thus, the reflectivity or albedo of these objects plays a role in their apparent magnitudes. To make Figure 22-1 it was assumed that all KBOs have an albedo of 0.1. Use the graph to determine the diameters of KBOs with absolute magnitudes of 2, 4, 6, 8 and 10. Likewise, determine the absolute magnitude of KBOs of 500, 100, and 40 km.
2. Table 22-1 is a list of the absolute magnitudes of all KBOs discovered in 2002. Use Figure 22-2 to convert each absolute magnitude to a diameter. Report the largest and smallest diameters measured. Report the average diameter measured and compare this average to the diameters of Pluto and Charon.
3. Complete Table 22-2 by filling the number of KBOs in each diameter range.

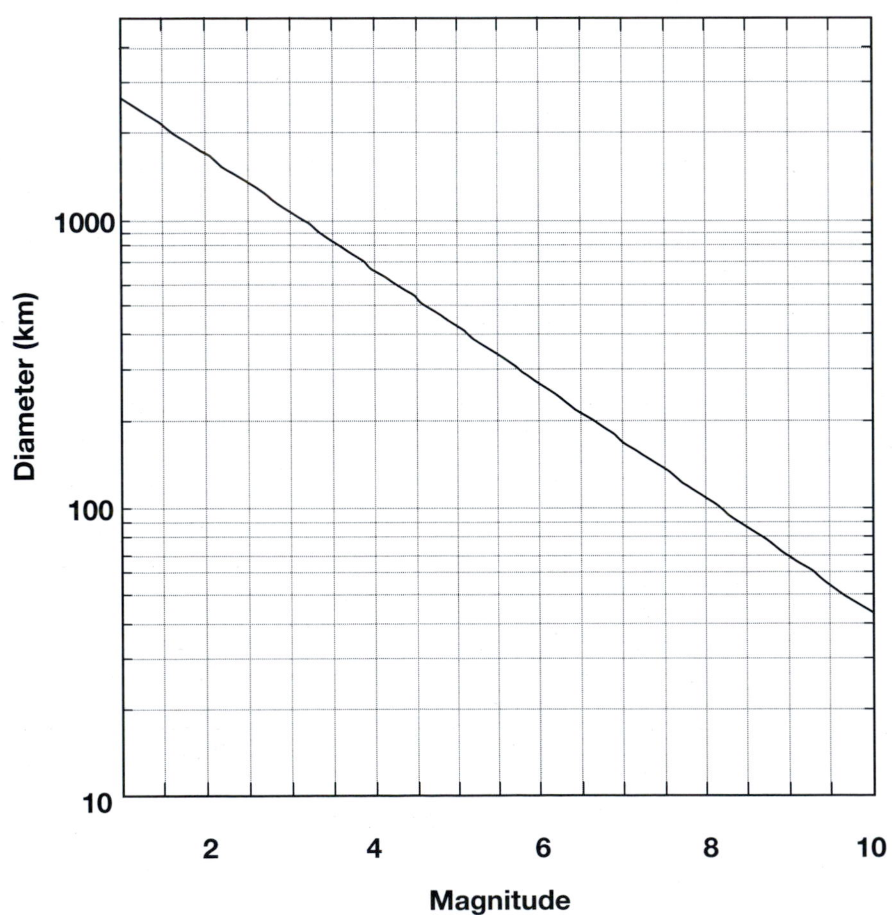

Figure 22-1. Relationship between diameter and absolute magnitude for KBOs, assuming an albedo of 0.1 for all KBOs.

4. Make a histogram by plotting the number of objects of a given size versus the size of the object, using the data from Table 22-2.

You are using the observed distribution of sizes to try to infer the true underlying distribution of sizes. You want to answer questions like: Does the number of objects increase or decrease with size? How rapidly? Does the distribution have a peak? With different data sets you could attempt to answer similar questions about how many total objects exist and where they are distributed in space.

Write a paragraph summarizing your results. Your paragraph should report results pertaining to the measured distribution (such as where the graph peaks) and what you might infer about the population of objects giving rise to the observed distribution (such as how the number of objects increases or decreases with size).

This type of work is an example of statistical astronomy. Other examples abound, particularly in the early stages of the study of newly discovered phenomena. In 1906 J.C. Kapteyn developed his *Plan of Selected Areas.* The plan called for measuring proper motions and apparent magnitudes for a vast number of stars. It was hoped one could then use the data to infer their distribution in space, or the structure of the galaxy. More recently, as astronomers worked to understand the nature of Gamma Ray Bursts in the 1980s and 1990s they made spatial maps of all known events and histogrammed durations of these events. Statistical tests were applied in an effort to find clues to what and where these objects are.

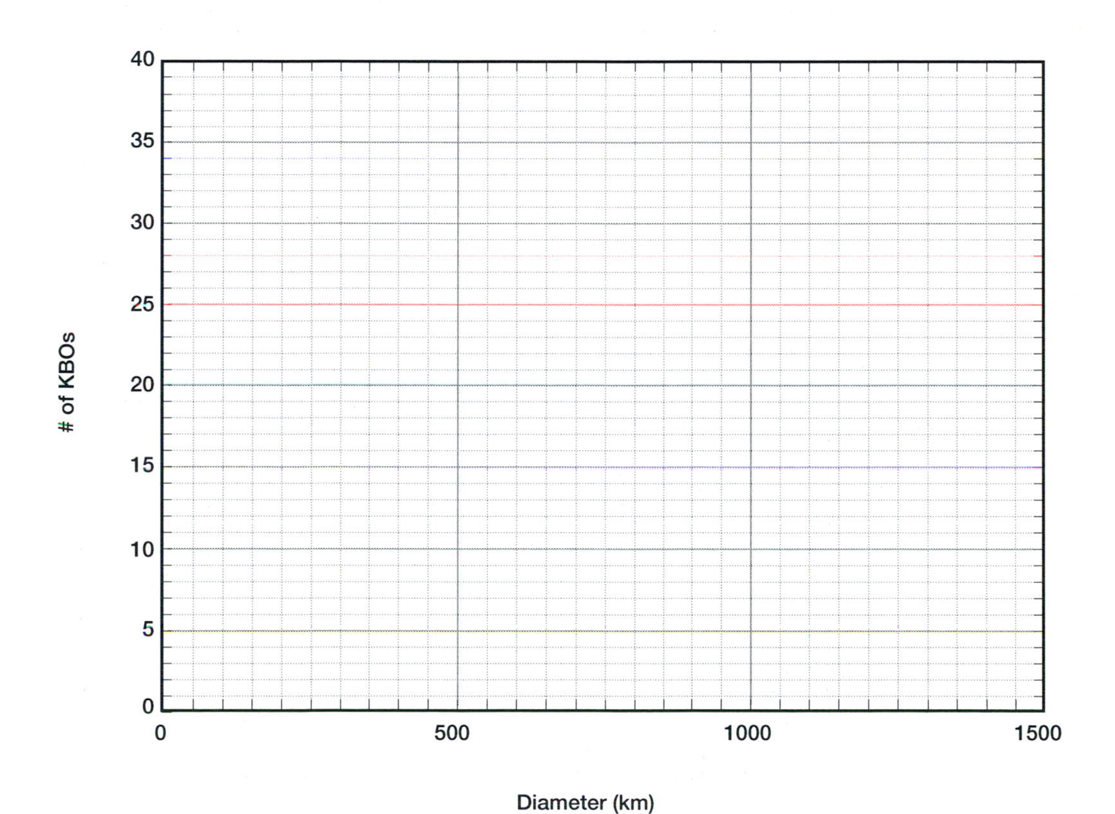

Figure 22-2.

Table 22-1.

The first column gives designation of each object. The absolute magnitude is found in the middle column and the date of discovery is found in the last column. All data is taken from the IAU list of KBOs (http://cfa-www.harvard.edu/cfa/ps/lists/TNOs.html).

2002 XW93	5.7	2002 12 10	2002 PG153	9.1	2002 08 15	2002 GJ32	5.6	2002 04 08
2002 XV93	4.5	2002 12 10	2002 PF153	9.0	2002 08 15	2002 GH32	5.6	2002 04 08
2002 XJ91	8.3	2002 12 05	2002 PE153	6.5	2002 08 15	2002 GF32	5.6	2002 04 08
2002 XH91	5.6	2002 12 04	2002 PD153	8.7	2002 08 15	2002 GE32	6.9	2002 04 08
2002 XG91	7.6	2002 12 04	2002 PC153	7.9	2002 08 15	2002 GD32	5.5	2002 04 07
2002 XF91	6.9	2002 12 04	2002 PB153	9.1	2002 08 15	2002 GC32	7.3	2002 04 07
2002 XE91	6.0	2002 12 04	2002 PA153	8.8	2002 08 15	2002 GZ31	6.2	2002 04 06
2002 XD91	7.6	2002 12 04	2002 PZ152	9.1	2002 08 15	2002 GY31	6.5	2002 04 06
2002 WC19	4.6	2002 11 16	2002 PY152	9.4	2002 08 15	2002 GX31	7.0	2002 04 06
2002 VF131	6.5	2002 11 07	2002 PX152	8.7	2002 08 13	2002 GW31	6.7	2002 04 06
2002 VE131	6.5	2002 11 07	2002 PW152	9.0	2002 08 12	2002 GV31	5.3	2002 04 06
2002 VD131	6.3	2002 11 07	2002 PV152	9.5	2002 08 12	2002 FX36	6.5	2002 03 18
2002 VC131	6.4	2002 11 07	2002 PU152	8.6	2002 08 12	2002 FW36	6.2	2002 03 18
2002 VB131	6.2	2002 11 07	2002 PT152	8.3	2002 08 12	2002 FP7	8.5	2002 03 22
2002 VA131	6.8	2002 11 09	2002 PS152	8.9	2002 08 12	2002 FX6	6.9	2002 03 20
2002 VY130	8.4	2002 11 07	2002 PR152	8.7	2002 08 12	2002 FW6	8.2	2002 03 20
2002 VX130	8.1	2002 11 07	2002 PG150	7.7	2002 08 11	2002 FV6	6.8	2002 03 20
2002 VW130	6.9	2002 11 07	2002 PQ149	6.1	2002 08 11	2002 FU6	7.3	2002 03 20
2002 VV130	7.5	2002 11 07	2002 PP149	6.4	2002 08 11	2002 CE251	8.4	2002 02 08
2002 VU130	5.7	2002 11 07	2002 PO149	5.8	2002 08 11	2002 CD251	7.2	2002 02 08
2002 VT130	5.6	2002 11 07	2002 PN149	7.0	2002 08 11	2002 CC251	8.0	2002 02 06
2002 VS130	6.3	2002 11 07	2002 PM149	5.6	2002 08 11	2002 CC249	6.5	2002 02 08
2002 VF130	7.1	2002 11 07	2002 PK149	7.0	2002 08 11	2002 CY248	4.9	2002 02 06
2002 VE130	6.2	2002 11 07	2002 PJ149	5.3	2002 08 11	2002 CB225	6.7	2002 02 07
2002 VD130	7.0	2002 11 07	2002 PH149	6.6	2002 08 11	2002 CA225	7.2	2002 02 07
2002 VR128	5.9	2002 11 03	2002 PF149	6.6	2002 08 11	2002 CZ224	6.5	2002 02 07
2002 VD95	8.6	2002 11 12	2002 PE149	6.6	2002 08 11	2002 CX224	6.0	2002 02 06
2002 VC95	8.0	2002 11 12	2002 PD149	5.8	2002 08 10	2002 CW224	6.9	2002 02 06
2002 VB95	8.9	2002 11 12	2002 PA149	6.0	2002 08 10	2002 CZ154	7.0	2002 02 06
2002 VA95	8.3	2002 11 12	2002 PN147	6.3	2002 08 09	2002 CY154	6.2	2002 02 06
2002 VZ94	7.1	2002 11 11	2002 PQ145	5.4	2002 08 09	2002 CW154	6.4	2002 02 06
2002 TM301	9.7	2002 10 06	2002 MS4	4.1	2002 06 18	2002 CV154	6.1	2002 02 06
2002 TL301	11.1	2002 10 06	2002 KX14	4.7	2002 05 17	2002 CU154	6.2	2002 02 06
2002 TJ301	10.9	2002 10 06	2002 KW14	5.7	2002 05 17	2002 CT154	6.6	2002 02 06
2002 TH301	10.7	2002 10 06	2002 GJ166	7.9	2002 04 09	2002 CS154	7.0	2002 02 06
2002 TG301	9.7	2002 10 06	2002 GH166	6.4	2002 04 09	2002 CR154	7.0	2002 02 06
2002 TF301	10.9	2002 10 06	2002 GG166	7.8	2002 04 09	2002 CQ154	6.4	2002 02 06
2002 TE301	10.9	2002 10 06	2002 GB33	7.8	2002 04 07	2002 CP154	6.0	2002 02 06
2002 TD301	10.6	2002 10 06	2002 GA33	8.6	2002 04 07	2002 CO154	6.5	2002 02 06
2002 TC301	9.5	2002 10 06	2002 GX32	7.0	2002 04 08	2002 VE95	5.3	2002 11 14
2002 TB301	9.1	2002 10 06	2002 GW32	6.5	2002 04 08	2002 UX25	3.6	2002 10 30
2002 TA301	10.2	2002 10 06	2002 GV32	6.5	2002 04 08	2002 TX300	3.3	2002 10 15
2002 TZ300	8.7	2002 10 06	2002 GU32	6.9	2002 04 08	2002 AW197	3.3	2002 01 10
2002 PE155	6.2	2002 08 12	2002 GT32	8.1	2002 04 08	2002 LM60	2.6	2002 06 04
2002 PD155	6.5	2002 08 12	2002 GS32	7.2	2002 04 08			
2002 PP153	7.3	2002 08 14	2002 GR32	7.8	2002 04 07			
2002 PO153	8.4	2002 08 14	2002 GQ32	7.7	2002 04 07			
2002 PN153	7.7	2002 08 14	2002 GO32	7.6	2002 04 06			
2002 PM153	9.0	2002 08 15	2002 GN32	6.3	2002 04 06			
2002 PL153	8.7	2002 08 14	2002 GM32	9.0	2002 04 06			
2002 PK153	7.7	2002 08 14	2002 GL32	7.6	2002 04 06			
2002 PJ153	8.9	2002 08 13	2002 GK32	6.4	2002 04 08			

Table 22-2.

Diameter (km)	Number of KBOs
0–50	
50–100	
100–150	
150–200	
200–250	
250–300	
300–350	
350–400	
400–450	
450–500	
500–550	
550–600	
600–650	
650–700	
700–750	
750–800	
800–850	
850–900	
900–950	
950–1000	
1000–1050	
1050–1100	
1100–1150	
1150–1200	
1200–1250	
1250–1300	
1300–1350	
1350–1400	
1400–1450	
1450–1500	

Discussion Questions

1. Look up the accepted albedo of Pluto. Compare it to the albedo assumed for KBOs in this activity. What would happen to the average size of the KBOs discovered in 2002 if they actually had an average albedo similar to that of Pluto?

2. At what diameter of KBO did your graph of number vs. diameter stop rising? Do you think that tells you more about the number of small KBOs or more about our ability to detect small KBOs? Explain.

3. Assume that the magnitude system used here varies the same way the stellar magnitude system varies (*i.e.*, $\Delta m = 1$ is factor of 2.512 in flux). If the smallest object discovered in 2002 is barely detectable at a distance of 35 AU about how far away would it be possible to detect the largest object discovered in 2002?

4. Suppose that some KBOs similar to the ones treated in this activity exist at distances of hundreds of AU from the sun. How might it be possible to detect these distant KBOs?

5. Notice that the KBOs in Table 22-1 were not discovered uniformly throughout the year. Can you suggest any reasons for this sporadic discovery rate?

6. How do the objects discovered on October 6 compare to those discovered the rest of the year? Discuss any difference. Does this inform the way you think about question 2?

Exercise

23

Determining the Velocity of a Comet

Purpose and Processes

The purpose of this exercise is to use two photographs of a comet, taken a short interval apart, to determine the velocity of the comet. The processes stressed in this exercise include:

Observing
Using Numbers
Interpreting Data
Inferring

Introduction

Comets have fascinated humankind for centuries. See Color Image 11 in the appendix. In ancient times these objects were regarded as omens of doom and the appearance of bright comets often shaped human events. For example, the appearance of a very bright comet in 1066 may well have influenced the outcome of the Battle of Hastings. The losing Saxons, under the leadership of Prince Harold, regarded the comet's appearance as a bad omen and may have been adversely affected. The bright comet of 1456 appeared at the same time as a Turkish invasion in western Europe and the two events were believed to be related. Pope Calixtus II ordered public prayers for deliverance from the evil effects of both the comet and the invaders.

Work done by Edmond Halley, in the early 18th century, showed that comets travel in orbits controlled by the gravity of the sun. Halley was able to show that some comets seen in the past were really the same comet returning at predictable intervals. In fact, he showed that the comets of 1066 and 1456 were the same comet. Examination of the movement of the comet convinced him that this comet should return to the vicinity of the sun at 75 year intervals. His successful prediction of its return in 1758 gave a powerful boost to the acceptance of Newtonian physics. The comet bears Halley's name today.

Comets are no longer regarded as bad omens. A large number of them have been discovered and their orbits charted in the last few centuries. Approximately 6 to 12 new comets are discovered each year and others are observed on their periodic return to the vicinity of Earth. Comets are usually named for their discoverers. If more than one individual makes an independent discovery, the comet is named for its first (up to three) codiscoverers. The comet you will investigate in this activity, Comet Kobayashi-Berger-Milon, was discovered in 1975.

Figures 23-1a and b are pictures of Comet Kobayashi-Berger-Milon. The original photographs were taken at the University of Northern Iowa. The exposures were 10 minutes long and were made with an SLR camera piggy-backed on a Celestron-8 telescope. A 50-mm lens was set at f/1.8 and Tri-X film was used. In one exposure, the stars were tracked and the comet was allowed to trail. In the second photograph, the comet itself was tracked. Can you tell which picture is which?

You will make measurements from these photographs to determine the position of the comet at the time of each exposure. From these positions it will be possible to calculate the angular distance that the comet moved in the two-hour interval between the pictures. Using this angular distance and the distance to the comet it will be possible to calculate the linear distance that the comet has moved in two hours. Once this linear distance is known, you will calculate the comet's velocity. In order to complete this activity, it might be useful for you to review the concept of plate scale as found in Exercise 17 *Orbit of the Moon.* The small angle formula should be reviewed as well. This formula relates the angular movement of an object in radians (θ) to its linear movement (s) and its distance (r)

$$s = r\,\theta.$$

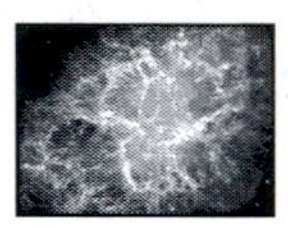

Procedure

1. Figure 23-1a was taken from 11:30-11:40 P.M. (CDT) on the night of July 24/25, 1975 and Figure 23-1b was taken two hours later. The fuzzy spot in the right central section of each picture is the comet. The brightest star in the field is Omicron Draconis, found to the left of the comet. Directly north of Omicron Draconis is SAO 029766. Figure 23-2 shows a section of a Smithsonian Astrophysical Observatory chart covering the same area of the sky. You will use the known coordinates of these two reference stars to establish a plate scale. You will then use your plate scale to determine the angular movement of the comet in the time interval between the two exposures.
2. Examine Figures 23-1a and b, and 23-2 and identify the comet and the two reference stars on each of the figures. Keep in mind that the scale of the drawing is not the same as the scale of the photographs. Differences in scale are a common problem in astronomical work and often require a very careful examination of visual records to insure that you are working with the correct celestial objects.
3. The coordinates of the two reference stars are as follows:

Star	*Right Ascension*	*Declination*
Omicron Draconis	$16^h\,0^m\,57^s$	58° 41′ 54″
SAO 029766	$16^h\,0^m\,58^s$	59° 46′ 04″

   (a) Find the difference in declination between the two stars in seconds of arc. Record this value on your data sheet. Remember that there are 60 minutes of arc in a degree and 60 seconds of arc per minute of arc.

   (b) Use a millimeter ruler and measure the linear distance between the images of Omicron Draconis and SAO 029766 on Figure 23-1a or b. Record this value to the tenth-millimeter.

   (c) The information obtained from the last two steps will be used to establish a plate scale. Divide the angular separation between the two reference stars in seconds of arc (Step 3a) by the linear photographic distance between the two reference stars (Step 3b). The result of this division is your plate scale for these photographs expressed as the number of ″/mm of arc. Record this plate scale on your data sheet.

Figure 23-1a. Comet Kobayashi-Berger-Milon (1975h) photographed from 11:30-11:40 P.M. (CDT) on the night of July 24/25, 1975.

Figure 23-1b. Comet Kobayashi-Berger-Milon (1975h) photographed from 1:30-1:40 A.M. (CDT) on the night of July 24/25, 1975. (Courtesy of Darrel Hoff and Tom Wagner, University of Northern Iowa.

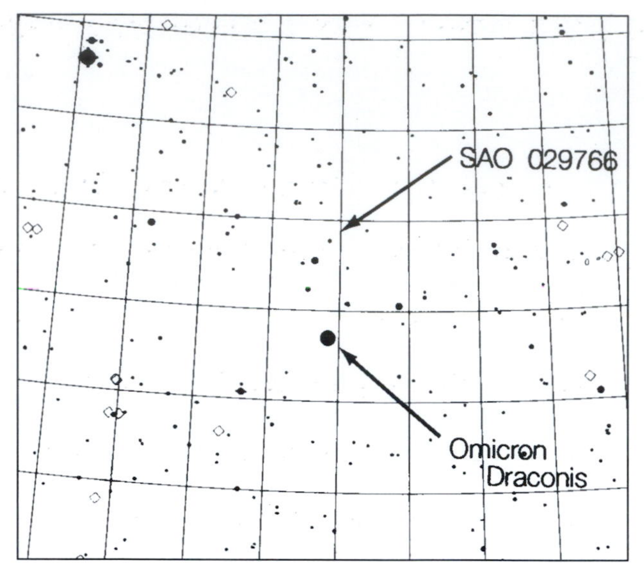

Figure 23-2. Section of SAO chart showing locations of Omicron Draconis and SAO 029766.

(d) Use a piece of tracing graph paper and mark the position of four or five of the brightest stars found on Figure 23-1a. Carefully mark the position of the comet on this tracing.

(e) Transfer this tracing to Figure 23-1b and align it with the same reference stars. When your reference stars are aligned with the star images on the second picture, carefully mark the second position of the comet image on your tracing paper.

4. (a) From your tracing, use a millimeter ruler to measure the distance the comet has moved in the two hours that elapsed between the two pictures. Record this measurement in your data table.

(b) Use the plate scale from Step 3c and convert the linear measurement from Step 4a into angular measurement. How many seconds of arc did the comet move across the sky in two hours? How many seconds of arc did it move in one hour? Record this hourly angular velocity in the data table.

5. On the night these photographs were taken, the comet was 0.609 astronomical units (A.U.) from the Earth, according to information supplied by the Central Bureau for Astronomical Telegrams at the Smithsonian Astrophysical Observatory. Convert this distance into kilometers. (Remember that there are 149,000,000 km/A.U.) Record this value in your data table.

6. During these exposures the comet was essentially moving from right to left across the sky and therefore its observed angular movement was a direct reflection of its linear movement. Use the hourly angular velocity of the comet from Step 4b, the distance to the comet from Step 5 and the small angle formula to calculate how far the comet moved in one hour. Record this value in your data table. Convert this velocity into km/sec by dividing by 3600.

Name ____ Date ______________________

Exercise 23. Determining the Velocity of a Comet

DATA SHEET

1. The difference in declination of Omicron Draconis and SAO 029766 is _________″.

2. The linear distance between Omicron Draconis and SAO 029766, as measured from Figure 23-1a, is _________ mm.

3. The plate scale of the photographs (linear distance/angular distance) is _________″/mm.

4. The distance between the comet images on the two photographs is _________ mm.

5. Using the plate scale (Answer 3 above), and the distance between the two comet images (Answer 4 above), the comet moved _________″ in two hours. Dividing this answer by two computes the comet's angular velocity as _________″/h.

6. The distance to the comet is 0.609 A.U. This distance converted into kilometers is _________ km.

7. Using the distance to the comet in kilometers and its calculated angular velocity, the calculated velocity of the comet is _________ km/h. Dividing this answer by the number of seconds in an hour (3600) computes the comet's velocity as _________ km/sec.

Use this space and the reverse side to show all calculations and to answer assigned discussion questions. Attach additional sheets if necessary.

Portrait of Fred Whipple

The late Fred Whipple of the Harvard-Smithsonian Center for Astrophysics in Cambridge, MA is known as the father of the 'dirty snowball' model for comets. Prior to his articulation of this model, comets were thought to be clouds of meteors surrounded at times by a gaseous envelope, but there was no universal agreement about the detailed structure of these perplexing objects. The Whipple model has come to be the accepted norm for comets today. During WWII Whipple was employed by the U.S. government on a project attempting to devise passive means of jamming German radar signals to confuse their antiaircraft weapons. This work led to the use of pathfinder planes dropping bundles of metal foil consisting of strips of measured lengths, which jammed the enemy radar beams. (Photo by D. Hoff.)

Portrait of Jan Oort

During the German occupation of Holland during World War II, Jan Oort devoted his time to the theoretical model of comet orbits and devised the current model of the origin of comets. This first satisfactory theory of comet origins was first published in 1950—the same year that Fred Whipple proposed the "snowball" model for the structure of the comet nucleus. In Oort's model he noted that in all cases where the orbits of new, nearly parabolic comets had been carefully determined, the orbits indicated a parhelion distance of approximately 50,000 times the Earth-Sun distance. Few comets came from closer distances and none showed evidence of coming from beyond the solar system. He then proposed that the existence of a comet cloud was associated with the sun, but existing beyond the farthest known planets. This gathering of comets is now known as the *Oort Cloud.* Most comets, therefore, are beyond the limits of detection, but occasionally, reasonably nearby stars can perturb the cloud and throw some comet nuclei in toward the center of our solar system. The comet nuclei then would brighten sufficiently to be observed from the Earth. Of course, random chance has other comets permanently thrown out of the solar system, never to be observed by humankind. (Photo by D. Hoff.)

Discussion Questions

1. What are some possible sources of error in this experiment?

2. At the time of the photographs, the comet was 0.53 A.U. from the sun. At this distance from the sun, the escape velocity from the solar system is 58 km/sec. How does this value compare with your measured velocity? What is the significance of this comparison?

PART V
STARS AND GALAXIES

Exercise

24

Proper Motion of a Star[1]

Purpose and Processes

The purpose of this exercise is to measure the proper motion of Barnard's Star and use this proper motion, with a given parallax and radial velocity, to determine the space motion of a star. The processes stressed in this exercise include:

Designing Experiments
Interpreting Data
Using Numbers
Using Logic
Inferring

Introduction

Humankind has long studied motion in the universe: our earliest naked eye observations of the movements of the Sun, Moon and planets led us to build models of the solar system and gave rise to modern astronomy. With the advent of the telescope, more subtle motions in the heavens became apparent. As early as 1718 Edmond Halley had noted that some stars were not fixed, but appeared to move in the sky relative to other stars. Arcturus in Boötes and Sirius in Canis Major were the first stars detected by Halley to have *proper motion*. This term is used to describe the yearly angular velocity of a star relative to a fixed field of stars, and is symbolized with the Greek letter μ. For example, Arcturus has a proper motion $\mu = 2.3''/\text{yr}$.

The star with the largest proper motion was discovered by E. E. Barnard in 1916 at Yerkes Observatory. This star, now called Barnard's Star, is a 9.5-magnitude star located in the constellation Ophiuchus. Its proper motion is so much larger than that of any other star that it is considered to be virtually a "runaway" star. Even so, the angular velocity is small enough that measurements must be made from photographs taken many years apart: the negatives of Barnard's Star in Figure 24-1 were taken in 1924 and 1951, respectively.

As proper motion is an angular velocity, we also need to know a star's distance to find its real velocity across our line of sight, called its *tangential velocity* (v_t). Two stars with the same proper motions can

[1]This exercise is adapted with permission from one developed by D. Scott Birney at Wellesley College.

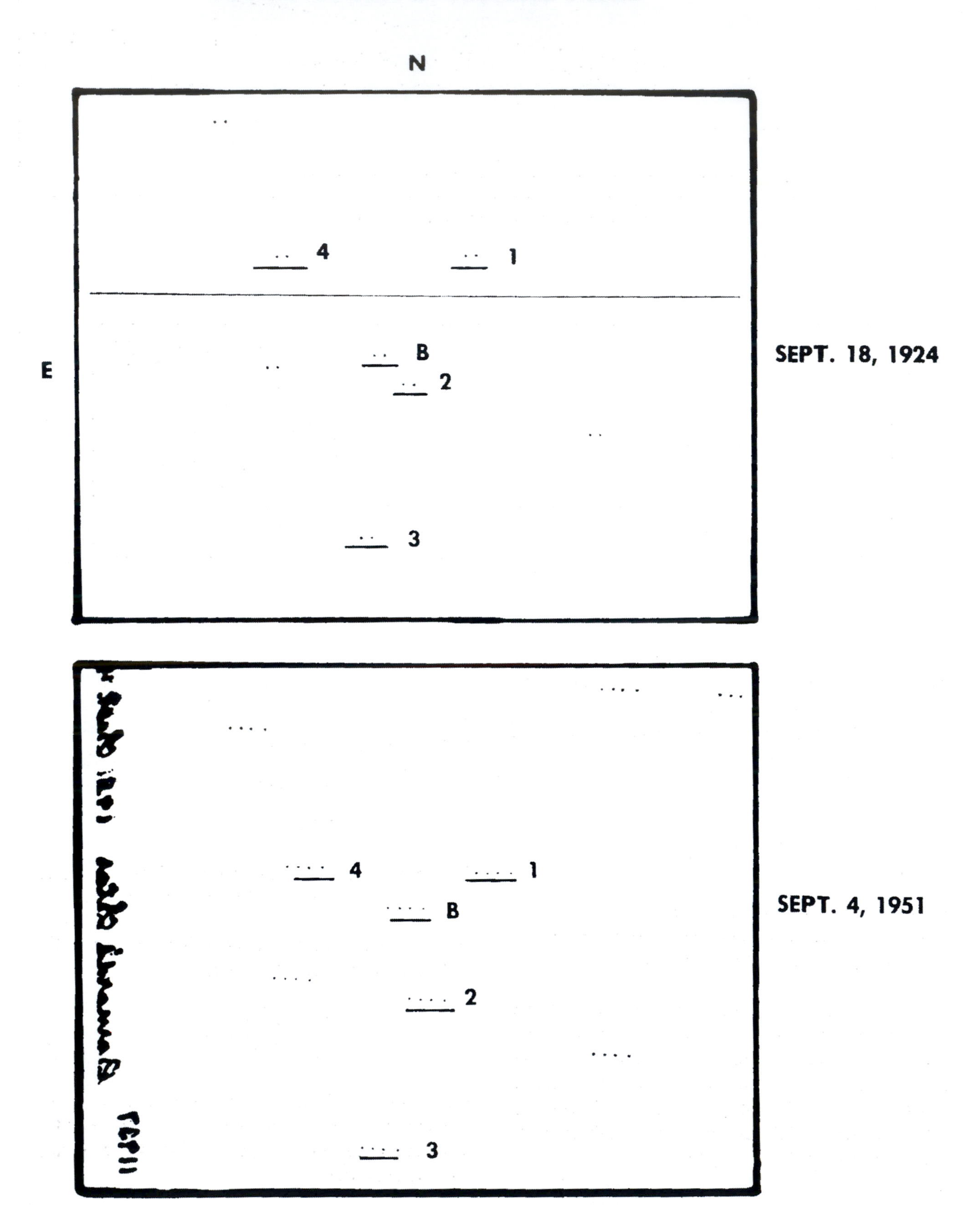

Figure 24-1. Proper motion of Barnard's Star.

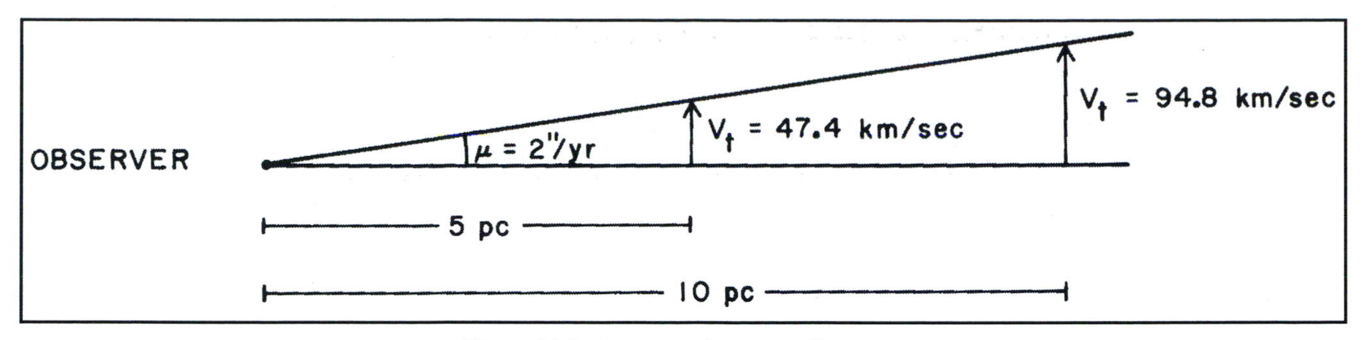

Figure 24-2. Proper motion at two distances.

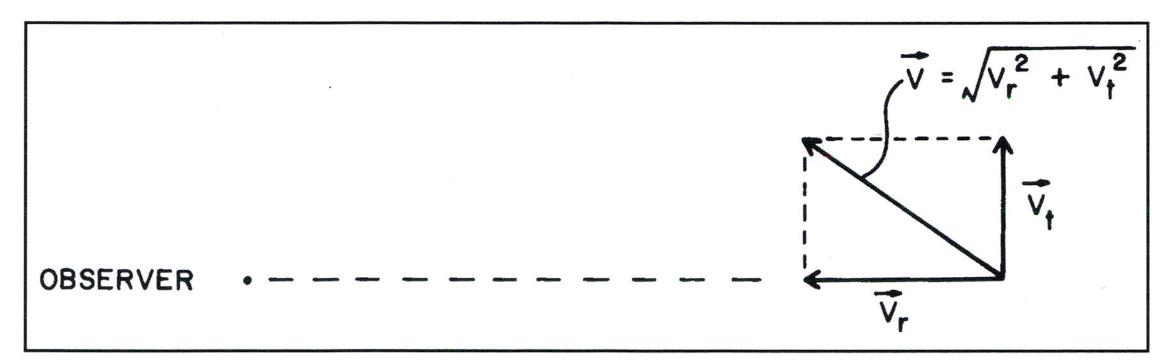

Figure 24-3. Space velocity.

have vastly different tangential velocities, as shown in the example in Figure 24-2.

Since stellar parallaxes are usually tabulated rather than stellar distances, tangential velocity can be calculated from the equation

$$v_t = \frac{4.74\ \mu}{p} \qquad (1)$$

where t = tangential velocity in km/sec
μ = proper motion in seconds of arc/year
p = parallax in seconds of arc.

We still need one more piece of information to know the velocity and direction a star moves in space, called its space motion or *space velocity.* So far we have considered only that part of a star's motion across our line of sight. The *radial velocity* (v_r), along our line of sight, is added vectorially (at right angles) to give us the total space velocity

$$\vec{v} = \vec{v}_r + \vec{v}_t$$

where $\vec{v}$ = space motion
$\vec{v}_r$ = radial velocity
$\vec{v}_t$ = tangential velocity

Remember that a negative radial velocity indicates motion toward us, and positive motion away.

The magnitude of v can be calculated using the Pythagorean theorem (Figure 25-3)

$$v^2 = v_r^2 + v_t^2 \quad ,$$
$$v = \sqrt{v_r^2 + v_t^2}. \qquad (2)$$

Space motions allow us to study the dynamics and interactions of groups of stars, and to see how the arrangement of our local group of stars is changing over the centuries.

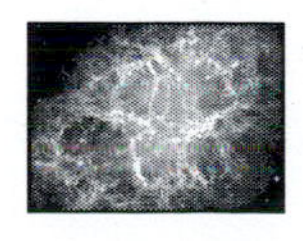

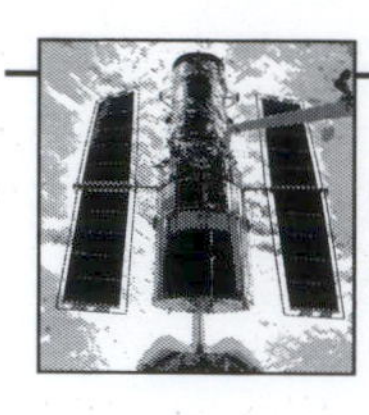

Procedure

PROPER MOTION

Figure 24-1 is a negative print of Barnard's Star and four labeled reference stars taken in 1924 and 1951.

1. Set up an arbitrary coordinate system on a piece of millimeter graph paper with the origin ($x = 0$, $y = 0$) near the lower left-hand corner. Align the graph paper over the 1924 negative so that the heavy horizontal line of the negative lays parallel to the x-axis. This establishes the east-west direction with east to the left and north toward the top of the negative. (A sheet of translucent graph paper is provided in Appendix 4.)
2. Using the coordinate system on the graph paper, trace and record the coordinates of the four reference stars and of Barnard's Star, marked *B*. Note that the stars were exposed more than once; use only the right-hand image of each star.
3. Place the graph paper over the 1951 negative and align it so the right-hand images of the four reference stars are superimposed on the plotted positions. Read and record the 1951 coordinates of Barnard's Star, and plot it on the graph paper.
4. Measure the distance between the two positions of Barnard's Star to 0.1 millimeter. Using a plate scale of 24.52 seconds of arc per millimeter, convert this motion to seconds of arc.
5. These two photographs were taken exactly 26.96 years apart. Calculate the proper motion of Barnard's Star in seconds of arc per year.

SPACE MOTION

1. The parallax of Barnard's Star has been measured to be 0.545″. Using Equation 1, find the tangential velocity of Barnard's Star in km/sec.
2. Barnard's Star has a radial velocity of -108 km/sec. Calculate the magnitude of the star's space velocity by using Equation 2 or by constructing a diagram similar to Figure 24-3.

CLOSEST APPROACH TO THE SUN

Since Barnard's Star's negative radial velocity indicates that it is moving generally toward us, we can find the epoch of its closest approach to the Sun and discover just how close that approach will be.

1. Draw a line to represent the direction of Barnard's Star on a piece of graph paper from a point representing the Sun. Calculate the distance to Barnard's Star in parsecs from the parallax-distance relationship ($d = 1/p$). Using a convenient scale (such as 10 cm = 1 pc), mark a point at the proper distance from the Sun to represent Barnard's Star.

2. Along the line from Step 1 and from Barnard's Star, draw an arrow or vector representing the radial velocity of the star (a scale of ½ mm = 1 km/sec works well). Again from Barnard's Star draw another vector (at right angles) representing the tangential velocity, using the same scale as for v_r. Complete the rectangle as in Figure 24-3 to find the direction and magnitude of the space velocity of Barnard's Star.
3. To find the point of Barnard's Star's closest approach to the sun, extend the diagonal of the space velocity to show the path the star will take. Using a compass or ruler, find the point along this line closest to the Sun. A line connecting the Sun and point of closest approach should be perpendicular to the path of Barnard's Star. Measure the closest approach distance in cm and convert to parsecs using the same scale factor as in Step 1 above. How close will Barnard's Star come to the Sun?
4. By again measuring from your diagram, determine how far (in pc) Barnard's Star has to travel from its present position to its point closest to the Sun. From the velocity and distance it has to travel we can calculate how long it will take to get nearest to us: velocity = distance/time, so time = distance/velocity. Using the conversion factor 1 km/sec = 1.02×10^{-4} pc/century, when will Barnard's Star be closest to the Sun?

Discussion Questions

1. At present our closest star is the Alpha Centauri system at about 1.33 pc. How will Barnard's Star compare to this at its closest approach?

2. How do we know that the four reference stars don't have proper motions of their own?

Exercise

25

Spectral Classification

Purpose and Processes

The purpose of this exercise is to become familiar with the classification system of stellar spectra. The processes stressed in this exercise include:

Observing
Classifying
Inferring

LABORATORY EXERCISES IN ASTRONOMY—SPECTRAL CLASSIFICATION

Auguste Comte, the French positivist philosopher, defined astronomy in 1835 as "the science by which we discover the laws of the geometrical and mechanical phenomena presented by the heavenly bodies." While the motions of the solar system could be analyzed and understood, the physical nature of the distant stars must be forever unknowable. "Men will never compass in their conceptions the whole of the stars," he stated.

Less than a quarter century later the German physicist Gustav Robert Kirchhoff laid down the laws of spectrum analysis, which opened a wholly new field for investigation. The study of the chemical and physical nature of the sun and stars, considered by Comte as hopelessly out of reach of the human intellect, became one of astronomy's most important branches.

After Kirchhoff's fundamental work, spectrum studies of the stars followed two different avenues. Astrophysicists such as William Huggins and Norman Lockyer combined laboratory experiments with detailed examinations of comparatively few stars, in an attempt to discover their composition and physical condition. On the other route, such pioneers as Angelo Secchi and E. C. Pickering undertook vast classifying schemes in order to understand better the diversity as well as the unity of the stellar populace. Ultimately both approaches were joined when Cecilia H. Payne (later Professor Payne-Gaposchkin) demonstrated that the apparent spectral differences arise primarily from variations in temperature or density, and that the overwhelming majority of stars in our galaxy must have quite similar chemical compositions.

Of all the great spectrum classification projects, none is more extensive than the famed *Henry Draper Catalogue,* a compendium of the spectral classes and magnitudes for 225,300 stars, prepared at Harvard Observatory in the early 1920s. (Later extensions to this catalogue

Owen Gingerich, *Smithsonian Astrophysical Observatory* (From Sky & Telescope.

include another 133,000 stars.) These classifications were made on photographs taken with medium-sized refractors. A thin prism covered the objective lens, producing a spectrum for each star on the plate.

Nowadays, Schmidt telescopes are generally used instead of refractors for taking objective-prism plates. The picture in Figure 25-2 was obtained with the 24-inch Burrell Schmidt and a four-degree objective prism at Warner and Swasey Observatory in Ohio. Case Western Reserve astronomers have recorded the blue and violet portions of the stellar spectra in one of the richest star clouds in the constellation of Cygnus, directly on the galactic equator. The field of some five square degrees lies just south of Gamma Cygni, the center of the Northern Cross.

Most of the properly exposed spectra in this reproduction come from 7th- to 9th-magnitude stars. The brightest is 5th-magnitude P Cygni, near the left (northern) edge of the picture, but its spectrum is quite heavily overexposed. Each spectrum would have been a narrow sliver of light if the telescope had been driven to follow the stars exactly, but instead the images were trailed in an east-west direction to widen them and make the spectral details easier to see and measure.

By close examination, we can distinguish the salient details of the Harvard system, which forms the basis of many more elaborate classification schemes in use today. This area of the sky exhibits a variety of spectral types, nearly every major class being represented. Particularly notable here are three of the 22 known Wolf-Rayet: stars of magnitude 8 and brighter.

This experiment consists of classifying the spectra of 30 stars. First we examine our enlarged samples on the key, where the stars are placed top to bottom in order of decreasing surface temperature. Violet is to the left, longer wavelengths to the right. Since the photograph was taken on a blue-sensitive plate, the red, orange, and yellow spectral regions have not been recorded, as already mentioned.

Note first that the hotter stars tend to be relatively much more intense in the violet than the cool ones. But it is to the narrow vertical lines in each spectrum that we look for making our detailed classification. This involves assigning a decimal subdivision within each of the major letter types. For example, *A*-type stars can be divided into 10 classes, *A*0, *A*1, *A*2, . . . , *A*9. Class *B*9 precedes *A*0, and *A*9 is followed by *F*0.

In general, the spectral lines are dark, caused by absorption of light in the star's atmosphere. But at the top of the key is one Wolf-Rayet object, the hottest star considered here, its spectrum containing bright (emission) lines. A quick glance at the bottom of the key might suggest that emission lines are found there too, but this is simply an illusion caused by the crowded pattern of dark lines in *K* and *M* stars. The increasing concentration of energy more and more to the long wavelengths is the first clue to these late-type stars.

In spectral types *B*, *A*, and *F*, the pattern of hydrogen lines (the Balmer series) plays an important role, immediately distinguishing these early types from the late ones. But the secret for distinguishing *B* and early *A* from late *A* and *F* is the relative strength of the K line of ionized calcium. The H and K pair becomes prominent in *G* and *K* spectra, and important secondary information is given by the strength of the G band at 4307 angstroms with respect to hydrogen-gamma (H_γ) or the neutral calcium line at 4227 angstroms.

Compare the following descriptions with the examples in the key chart:

- *O*—Temperature is so high in this class that helium is singly ionized and other elements are at least doubly ionized. In the visual region these spectra are almost featureless, but strong in the ultraviolet.

- *B*0—On a long continuum the Balmer series of hydrogen is faintly visible; if the spectrum is well exposed, a few helium lines may be seen. Neutral helium is strongest at *B*2, and then fades rapidly toward *A*0.
- *A*0—Hydrogen lines of the Balmer series are strong (strongest at *A*3); helium is no longer present. On this reproduction the calcium K line is faintly visible in the *A*2 star.
- *F*0—The Balmer lines are still conspicuous, although only half as strong as in *A*0; the K line of ionized calcium is as strong as the blend of H_ϵ (hydrogen-epsilon) and the H line of calcium.
- *G*0—In this solar-type spectrum, the H and K lines are the strongest features, with the Balmer lines no longer conspicuous. The continuum spectrum shows through between the numerous metal lines that are just at the limit of visibility, sometimes giving a false impression of emission lines. This aspect is even more troublesome in the *K* stars.
- *K*0—The energy maximum of the spectrum lies far to the red of the ionized calcium H and K lines, which reach their greatest intensity in this class. Many metal lines are easily visible. The strongest is that of neutral calcium at 4227 angstroms. Even stronger is the G band 80 angstroms to the red of the calcium line.
- *M*0—The wide bands of TiO, shaded toward the violet, mark the spectra of the *M* class. The 4227 line of calcium is very strong, and the G band is also conspicuous.

We now investigate the spectra on the Cygnus photograph itself. As a trial, sort the first six numbered spectra near the top into some rough order. Notice that 1 and 5 have similar, almost featureless, continua; 2 and 6 are related, both exhibiting many hydrogen lines; 3 is an emission-line star; 4 belongs in yet another category. The intense part of 4 is very short and is concentrated on the redward end. This indicates a much cooler star than 5, whose spectrum is very strong in ultraviolet light. It is evident that the longer spectra, such as 5, arise from hot, early-type stars, whereas 4 represents a late-type star.

Now compare 2 and 6 more carefully. The hydrogen lines are comparatively stronger in 2, but the most striking difference is the appearance of the H and K pair in 6, which spoils the regular pattern of the Balmer series. Comparison with the key shows that 2 is an *A* star and 6 an early *F,* in fact *A*0 and *F*2, respectively.

In 4 the H and K lines are clearly visible, but the hydrogen series has vanished, thus ruling out type *G* and earlier. The dark line some distance to the right of the H and K pair is the G band. Since the 4227 calcium line is not visible, this must be an early *K,* actually *K*2. It is instructive to compare this spectrum with 26, 27, and 28, as well as with the key.

Examples 1 and 5 are clearly much earlier than *A*0, belonging in the *O-B* classes, but an exact classification is difficult on the reproduction because the critical absorption lines are very fine. The long smooth spectrum 5, with no trace of hydrogen lines, reveals an *O* star. In 1 the Balmer series is faintly visible, indicating a *B*0 star.

Armed with the key photograph and accompanying description, as well as the examples above, you should be able to classify most of the 30 marked stars on the plate to within two or three decimal subclasses. Watch out for 9, however, for it is two almost superimposed spectra of the same type, and for 10, which has two overlapping parts—classify only the star to the left.

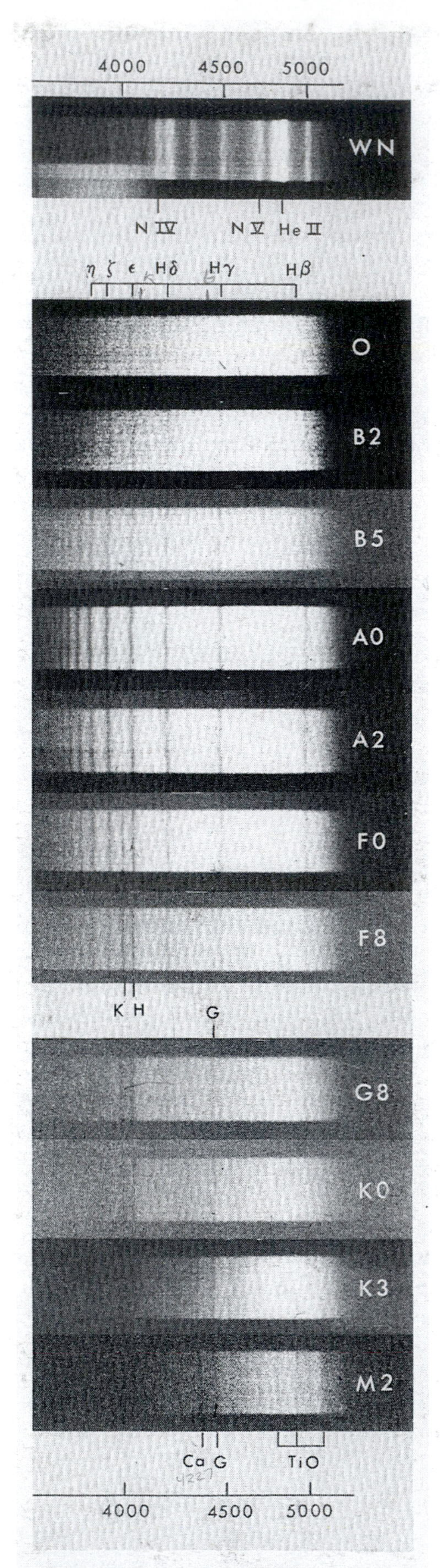

Compare these typical spectra with the detailed descriptions on the next page. Wavelengths are labeled in angstroms.

Figure 25-1. (Left) Compare these typical spectra with the detailed descriptions. Wavelengths are labeled in angstroms. (Warner & Swasey Observatory, Case Western Reserve University. Reprinted by permission.)

Figure 25-2. (Facing page) An objective-prism spectrogram of a region near Gamma Cygni, taken October 16-17, 1950, with the 24-inch Schmidt telescope of Warner and Swasey Observatory. The exposure was four minutes on Eastman IIa-O blue-sensitive emulsion, by Daniel Harris, III.

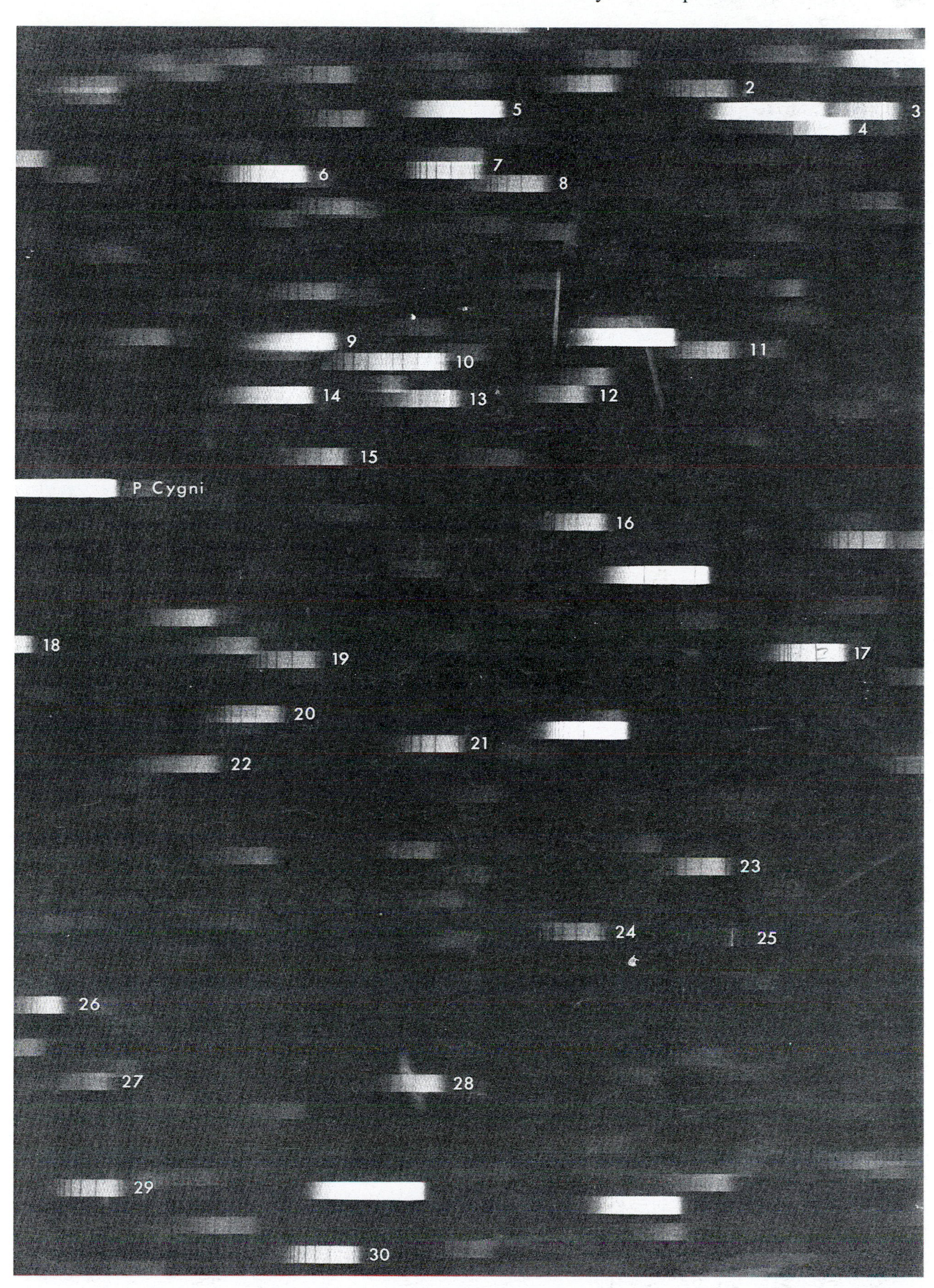
2
5
3
4
6
7
8
9
10
11
14
13
12
15
P Cygni
16
18
19
17
20
21
22
23
24
25
26
27
28
29
30

Portrait of Owen Gingerich

Dr. Owen Gingerich is acknowledged to be the premier historian of astronomy in the United States and is internationally recognized for his efforts. Owen is an Iowa-born astronomer and has been associated with Harvard University nearly all of his professional life. In addition to his work on the history of astronomy he has been a strong advocate for the improvement of astronomy education at all levels. For years he has served as the consulting editor of a series of articles published in the popular journal, *Sky and Telescope*, from which this laboratory exercise on spectral classification is reprinted with permission. (Photo by Darrel Hoff.)

Name ______________________ Date ______________

Exercise 25. Spectral Classification

DATA SHEET

1. My estimates for the spectral type of stars 1–5 are:

   1. ______
   2. A 0 class
   3. Wolf-Rayet
   4. K 2
   5. O class

2. My estimates for the spectral type of stars 6–25 are:

   6. F2
   7. A0
   8. A2
   9. O
   10. ______
   11. ______
   12. ______
   13. ______
   14. ______
   15. ______
   16. ______
   17. F
   18. ______
   19. ______
   20. ______
   21. ______
   22. ______
   23. ______
   24. ______
   25. ______

Discussion Question

1. Notice that the spectra of hotter stars are brighter on the left end (shorter wavelengths) than on the right side. Notice that the opposite is true of the cooler stars. What explanation can you give for this?

Exercise 26

A Color-Magnitude Diagram of the Pleiades[1]

Purpose and Processes

The purpose of this exercise is to construct and examine a color-magnitude diagram of the Pleiades star cluster. The processes stressed in this exercise include:

Using Numbers
Interpreting Data
Formulating Models

Introduction

As astronomers developed techniques for studying some of the physical properties of the stars, they also searched for relationships among these properties. Two of the most easily measured properties of any star are its apparent magnitude, measured either photographically or photoelectrically, and its temperature, determined from its stellar spectral class or *color index.*

Color index is a comparison of the amounts of blue and yellow (or visual) light a star radiates: a hot star radiates more blue and violet light, while a cooler one radiates more yellow and red light (Figure 26-1).

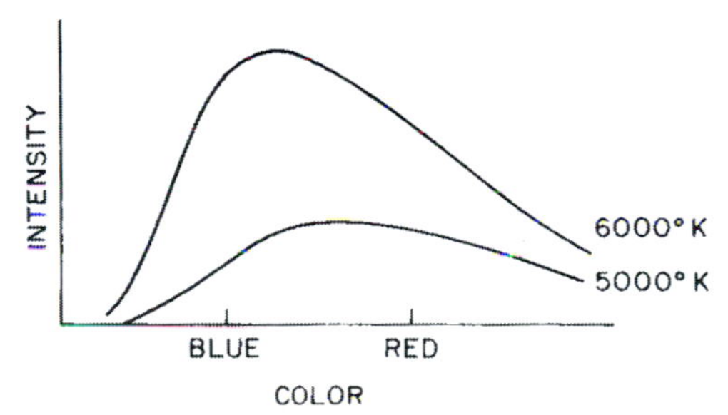

Figure 26-1. Radiation curves.

Color index can be defined then as

$$CI = m_b - m_v$$

[1]This exercise is adapted from *Astronomy: Observational Activities and Experiments* by Michael K. Gainer. Copyright © 1974 by Michael K. Gainer. Reprinted by permission of the author.

and is often written

$$CI = B - V$$

where CI is the color index
$m_B = B$ is the apparent blue magnitude
$m_v = V$ is the apparent visual magnitude.

Apparent stellar magnitudes in the blue spectral region can easily be measured photographically because most film is more sensitive to blue light. Visual magnitudes are measured from photographs taken with an amber filter. Standardized blue, visual, and ultraviolet filters have been developed for photoelectric use.

In using the color index of a star we must remember the way the stellar magnitude scale is set up: large magnitude numbers refer to dimmer stars. Thus, for a hot blue star, B will be numerically less than V (indicating more blue than visual light), and the color index will be a negative number. Likewise, cooler stars have positive color indices.

Because a star's apparent magnitude depends on its distance as well as its intrinsic brightness, we really would not expect a correlation between apparent magnitude and temperature (or color index) for most stars in the sky. However, it might seem reasonable for a star's temperature to be related to the total amount of energy it radiates, or its absolute magnitude. Unfortunately, absolute magnitudes are hard to determine directly because of the difficulty in determining reliable distances for more than a few hundred stars in the sky.

One way to avoid this difficulty is to study the magnitudes and colors of stars in a cluster. All stars in the cluster are at approximately the same distance, so the apparent magnitude of each differs from its absolute magnitude by the same factor. By comparing color index and apparent magnitude we are also relating temperature and absolute magnitude.

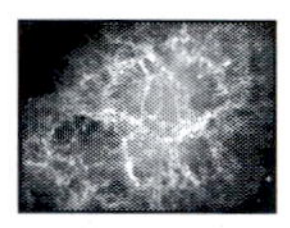

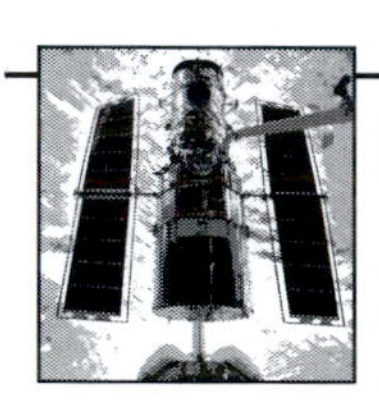

Procedure

Figures 26-2a and b are photographs of the Pleiades taken without and with an amber filter, respectively. A 135-mm focal length lens was used with Tri-X film at f/2.8, with exposure times of 1.0 and 1.5 minutes. (A longer exposure time must be used with a filter to compensate for the absorption of light by the filter.)

1. Try to obtain your own photographs of the Pleiades if possible. Lenses with focal lengths from 100 mm to 200 mm work well at the lowest f-ratio setting, with exposure times ranging from one to three or four minutes. Allow approximately one and one-half times as long an exposure time for the photograph with the filter. To minimize the effects of possibly changing atmospheric conditions, alternate frames taken with and without the filter rather than taking a series with the filter followed by a series taken without.
2. Figure 26-3 is a comparison scale to use in estimating stellar brightnesses. Number the images from 1 to 9, starting with the largest. Although these numbers do not correspond to the standard stellar magnitude scale, they are proportional to magnitudes. Cut out the comparison scale so that it can be placed adjacent to the star images in Figures 26-2a and b.
3. Measure the magnitude of a star on the print for which no filter was used by comparing the size of the star image with the comparison scale. If a star image falls between two images on the scale, e.g., between 4 and 5, assign it a value of 4.5.

   After you measure a particular star on the print for which no filter was used, move immediately to the print obtained with the amber filter and measure the same star. Number each star after you measure it to avoid measuring the same star twice.

Figure 26-2a. A photograph of the Pleiades taken without a filter.

Figure 26-2b. A photograph of the Pleiades taken through an 85B amber filter. (Courtesy of Michael K. Gainer, St. Vincent College.)

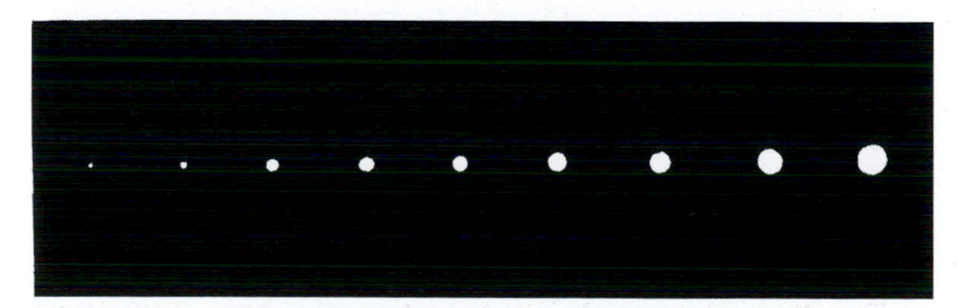

Figure 26-3. Magnitude comparison scale.

4. Measure at least 20 stars, selecting samples of all magnitudes. Select stars only from around the center of the print. Do not measure stars which cannot be distinguished as single stars.
5. Plot a color-magnitude diagram of B (from the photograph taken without a filter) versus B-V, the difference between the two magnitudes.
6. Discuss any conclusions you can make from this graph.

Name ______________________ Date ______________

Exercise 26. A Color-Magnitude Diagram of the Pleiades

B

B-V

Discussion Question

1. How do you interpret data that fall well outside the majority of the points on your plot?

Exercise

27

Supernova 1987 A

Purpose and Processes

The purpose of this exercise is to determine the absolute magnitude of a supernova when it is at peak brightness using measured image diameters on photographic plates. The processes stressed in this exercise include:

- **Using Numbers**
- **Interpreting Data**
- **Inferring**
- **Controlling Variables**

Introduction

Supernovae are spectacular explosive events that result from core collapse in the final stages of stellar existence. The events are relatively rare but have been studied extensively in recent years. One reason for the increased scrutiny of supernovae is that Type I supernovae (see below) appear to be very good "standard candles" that can be used to effectively determine the distances to the most remote galaxies. This distance determination allows astronomers to measure the expansion rate of the universe in different epochs. See Exercises 11 and 14. Spiral galaxies, like the Milky Way, appear to produce supernovae at the rate of a few per century. These events can be seen in distant galaxies but most in our own galaxy are completely obscured by dust. The last visible supernova in the Milky Way occurred in 1667. The local supernova rate has been estimated by the measuring gamma ray emission from radioactive nuclei produced in supernovae and the local rate appears roughly consistent with what is observed in other spiral galaxies.

Supernovae are commonly divided into two classes by astronomers, Type I and Type II. Type I supernovae brighten to a peak magnitude of about -19 in just a few days. They fade very quickly, for a few days, then more slowly over a period of a year or more. Type II supernovae are not as bright at peak brightness and they do not fade in as regular a fashion as Type I supernovae. The differences between the two types are caused by the differences in the pre-explosion stars. Type I supernovae are believed to occur in binary systems in which a star loses material onto the surface of its white-dwarf companion star. The resulting extra gravitational pressure causes the white dwarf to collapse and produces the supernova explosion. A Type II supernova occurs when a massive single star collapses after photodisintegration of its iron core.

The Large Magellanic Cloud (LMC), a small irregular galaxy, is a satellite of the Milky Way and is located about 50 kiloparsecs away. Its name attests to the fact that it was discovered by Magellan on his historic voyage around the world. On February 23, 1987 astronomers witnessed a naked-eye supernova that came to be known as Supernova 1987A (SN1987 A). The name denotes that it was the first supernova observed that year. The occurrence of a supernova this close to Earth ranks as one of the most exciting astronomical events in recent years.

The placement and timing of SN 1987A was very fortuitous. It was located in a galaxy whose distance is accurately known. It was observed almost from the beginning of the explosion and (unlike all previous supernovae) information was known about the progenitor star. One disadvantage was that the event could only be seen from southerly latitudes. The United Kingdom 1.2-m Schmidt Telescope, located at Coonabarabran, New South Wales, Australia produced an excellent series of photographs showing the changing brightness of the star. A portion of that photographic series is shown in Figures 27-2, 27-3, and 27-4. These negative prints also show the bright nebula, 30 Doradus, up and to the left of the supernova. Table 27-1. lists the dates on which the photographs were taken.

As SN 1987A changed in brightness, the diameter of the image of the star changed on photographic plates. This change in image size is proportional to the change in the *brightness* of the star and should not be interpreted as a change in the diameter of the star itself. (See Exercise 26 *A Color-Magnitude Diagram of the Pleiades* for another example of the use of photographic photometry.) The diameter of stellar images can be measured and a light curve constructed. A light curve consists of a plot of the brightness of an astronomical object plotted against the time at which each measurement was made. Light curves reveal information in a graphical form that assist astronomers in interpreting the behavior of variable objects. You will use the light curve to determine the apparent magnitude of SN 1987A at its brightest. From that information you will determine the absolute magnitude of the star at peak brightness.

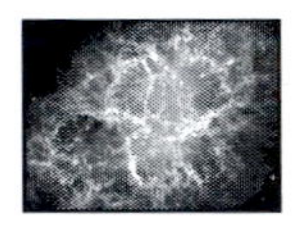

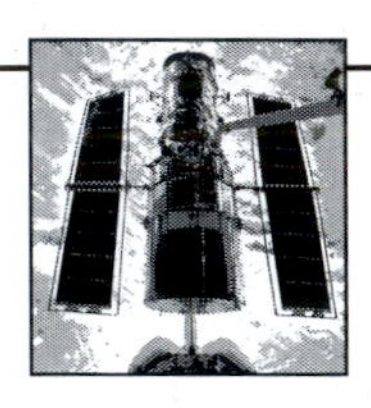

Procedure

1. Examine the accompanying data table (Table 27-1). Notice that for each observation date, column three shows the elapsed time since the start of the supernova event on February 23, 1987.

Table 27-1. Photograph Dates

Photograph	Date	Days Since Event
a	27 February 1987	4
b	10 March	15
c	29 March	34
d	8 April	44
e	24 April	60
f	5 May	71
g	25 May	91
h	4 July	131
i	15 August	173
j	17 October	236
k	9 December	289
l	11 February 1988	353

2. Use either a metric ruler or vernier caliper to measure the supernova image diameter (to the nearest 0.1 mm) on each photograph at least twice and average your results. Record this average in your data table. Note that the brightest images of the supernova are surrounded by a halo. This halo is not connected to the star in any way and is only an artifact of the photographic process. By carefully examining each image, you should be able to distinguish the true image size from the surrounding halo.
3. On the graph paper provided, plot each of the SN 1987A image diameters versus the number of days since the supernova exploded. Draw a smooth curve through the plotted points. The date of maximum brightness may not have been recorded because of the timing of the photographs. Use your smoothed curve to estimate the date peak brightness occurred.
4. The usual way to determine the photographic magnitude of a star is to locate a number of stars of known magnitudes in the photographic field and measure their image diameters. The magnitude of a star can be estimated by comparing its diameter to the image diameters of stars of known magnitude. For the SN 1987A photographs provided here, the field does not contain any star as bright as the supernova at its peak brightness.
   Figure 27-1 is provided to assist in estimating the peak magnitude of SN 1987A. It shows the image sizes that stars of known magnitude would have if they were photographed with the same system used to photograph the supernova.
5. Use your graph to determine the supernova's image diameter at peak brightness. Measure the diameters of the representative dots in Figure 27-1 and find the one that is closest in size to the supernova image at peak brightness. Estimate the apparent magnitude (m) of the supernova at its brightest.

Did You Know?

As optically bright as a supernova is, a much greater portion of its energy is released in the form of neutrinos. Neutrinos are particles that are produced in certain types of nuclear reactions. They don't interact readily and are, thus, difficult to detect. More than ten trillion neutrinos pass through the detector each second. For many years, large underground detectors have been used to detect neutrinos that are produced in the nuclear reactions powering the Sun. Two such detectors, which were actually built to look for proton decay, saw neutrinos from Supernova 1987 A, representing the first time neutrinos had ever been detected from a known source beyond the solar system. The Kamiokande II detector in Japan used 3,000 metric tons of water as a detecting medium; it detected 11 neutrinos from the supernova. The Irvin-Michigan-Brookhaven (IMB) detector in the U.S. used 8,000 tons of water to detect 8 neutrinos from SN 1987 A. This advent of extra-solar neutrino astronomy allowed investigators to test supernova models. Shortly after the event scientists lamented the fact that the detector could have been built better and detected far more neutrinos from the explosion. Unfortunately theorists had calculated that IMB would not be able to see neutrinos from a supernova and it was not worth the added expense of improving the detector which, recall, was built for a proton decay experiment. The added neutrinos would have been useful in understanding both supernovae and neutrino physics. The episode highlights a dilemma of modern astronomy. In general, one would like to build the best detector possible because one never knows what surprise might be found, but the reality of limited resources leads to compromises in the building of telescopes and detectors.

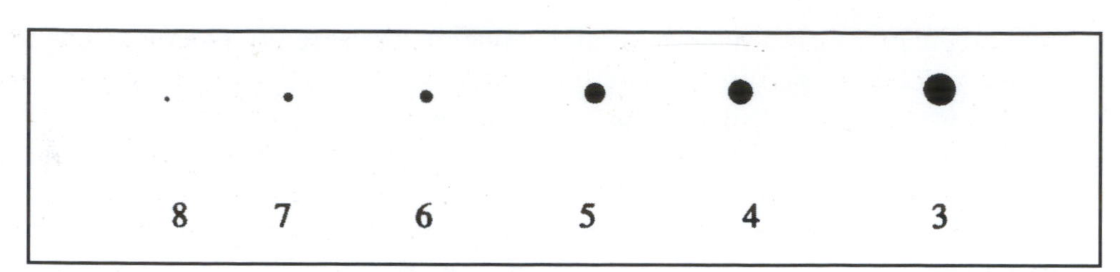

Figure 27-1. Reference stellar images. (The number below each image represents the magnitude of that image.)

6. Calculate the absolute magnitude (M) of SN 1987A using the formula

$$M = m + 5 - 5 \log r$$

where m is your estimated peak apparent magnitude and r = 50,000 pc, the known distance to the LMC.

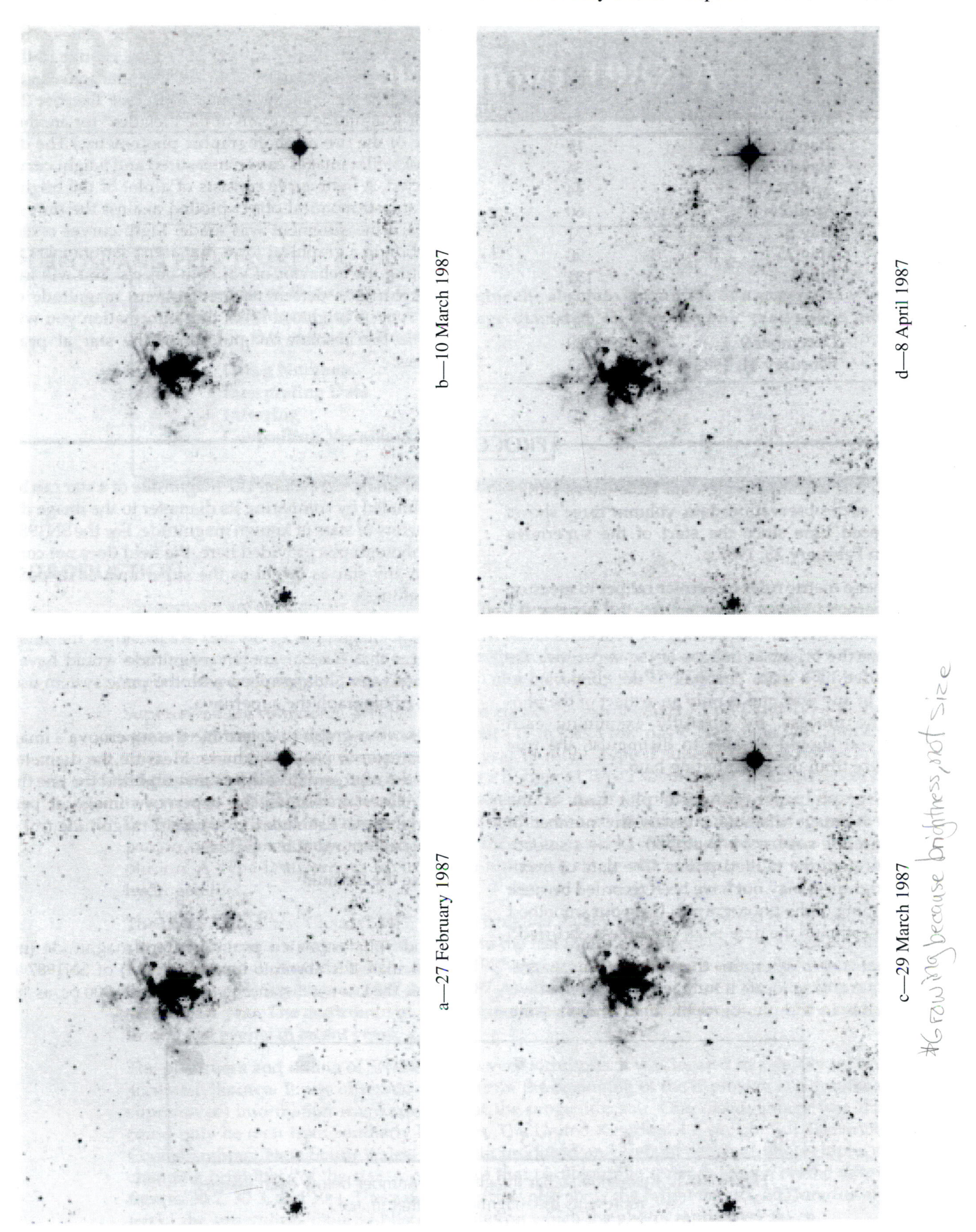

Figure 27-2. Supernova 1987 A in the LMC. (Copyright © PPARC, Royal Observatory, Edinburgh.)

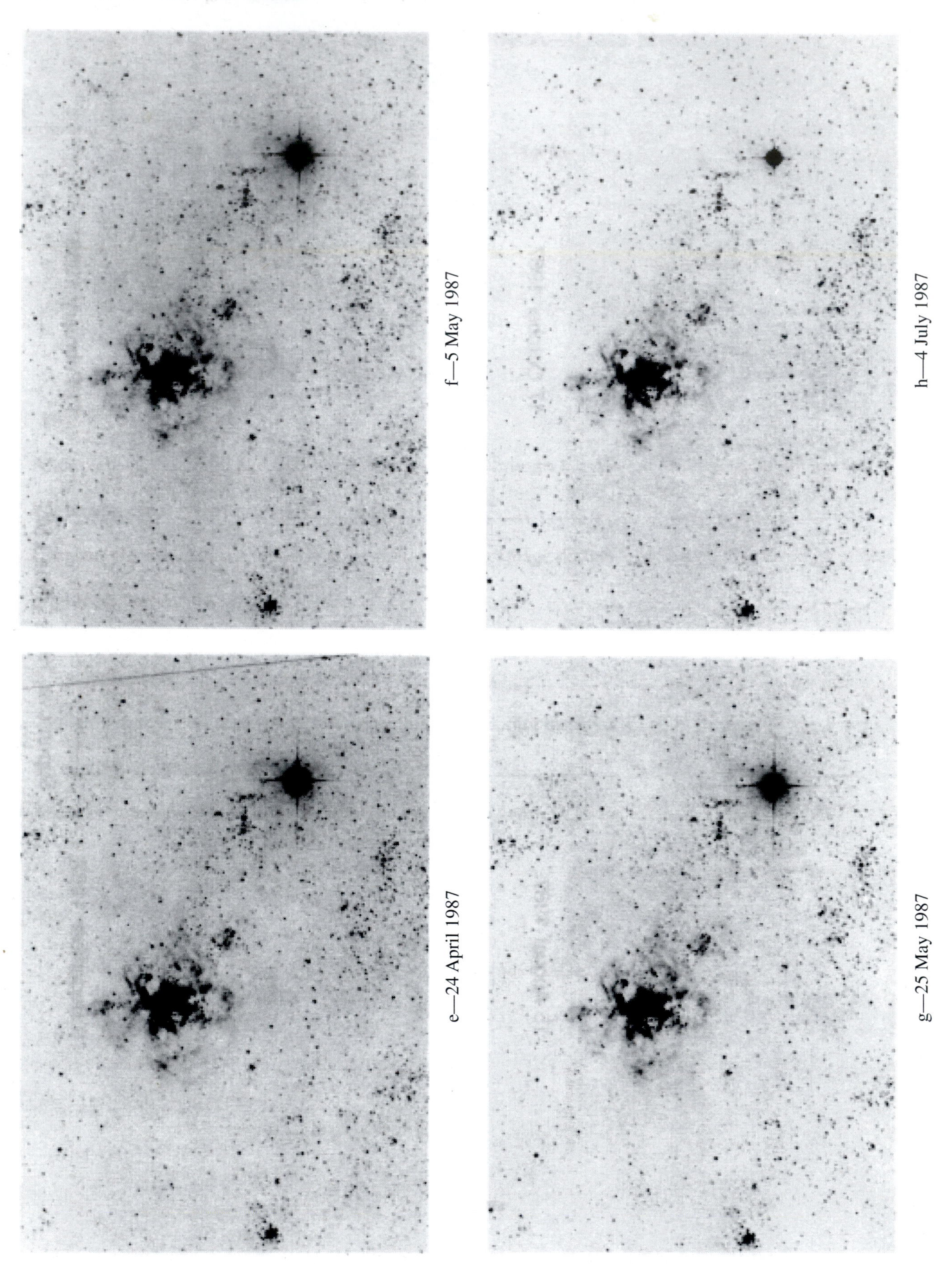

Figure 27-3. Supernova 1987 A in the LMC. (Copyright © PPARC, Royal Observatory, Edinburgh.)

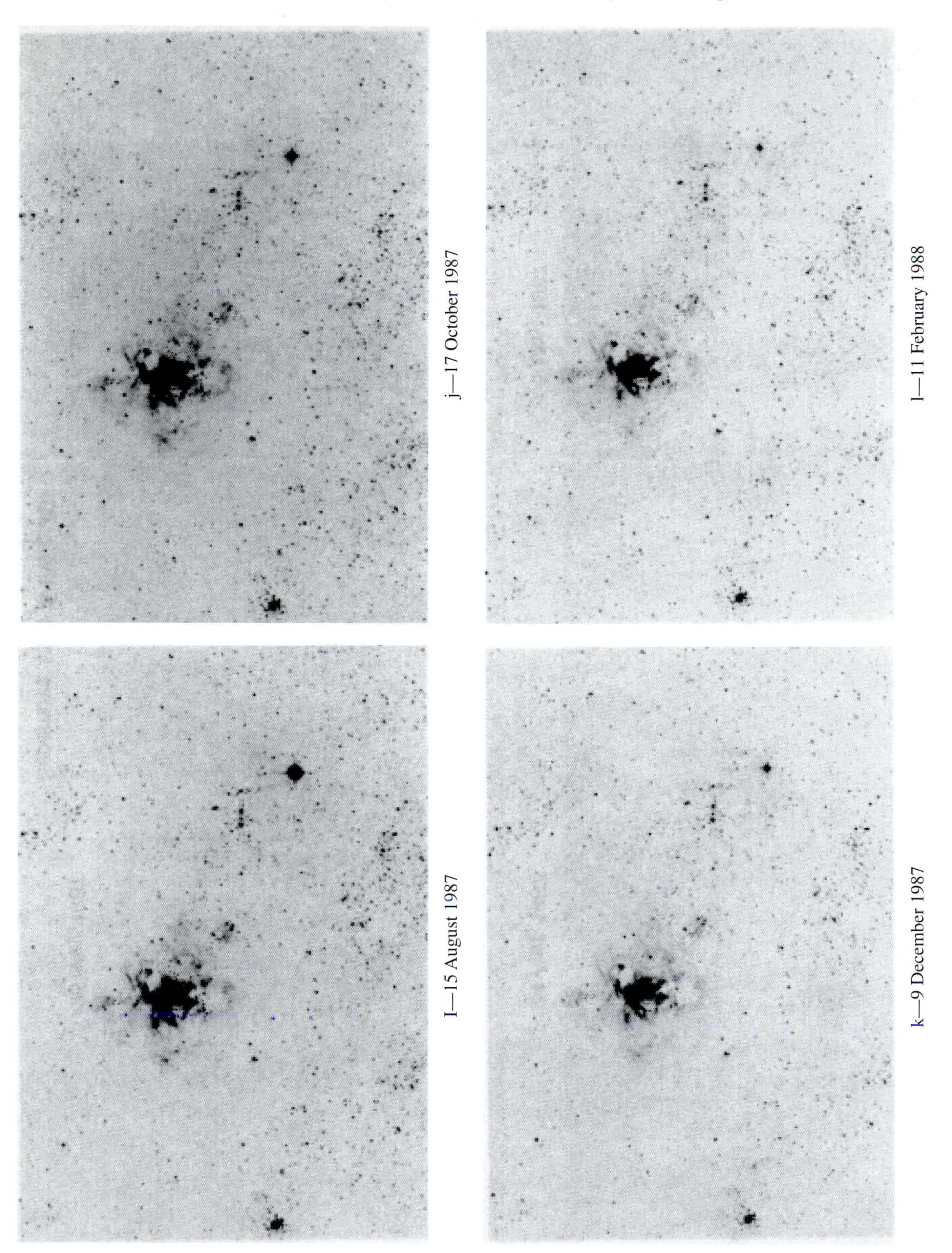

Figure 27-4. Supernova 1987 A in the LMC. (Copyright © PPARC, Royal Observatory, Edinburgh.)

Name ________________________ Date ________________

Exercise 27. Supernova 1987 A

DATA SHEET

Photograph	Date	Days Since Event	Image Diameter (mm) 1st Measure	2nd Measure	Average
a	February 27, 1987	4	3.0 mm	3.1 mm	3.05 mm
b	March 10	15	3.5 mm	3.4 mm	3.45 mm
c	March 29	34	3.8 mm	3.6 mm	3.7 mm
d	April 8	44	3.9 mm	3.9 mm	3.9 mm
e	April 24	60	6.0 mm	5.9 mm	5.95 mm
f	May 5	71	6.0 mm	5.7 mm	5.85 mm
g	May 25	91	5.7 mm	5.7 mm	5.7 mm
h	July 4	131	3.2 mm	3.1 mm	3.15 mm
i	August 15	173	2.8 mm	2.9 mm	2.85 mm
j	October 17	236	1.8 mm	2.0 mm	1.9 mm
k	December 9	289	1.0 mm	0.9 mm	0.95 mm
l	February 11, 1988	353	0.8 mm	0.9 mm	0.85 mm

1. After plotting the graph and drawing a smooth line through the data points, determine the date of maximum brightness. What is that date? May 3

2. Use your smoothed curve to arrive at an estimate of the image diameter of SN1987 A at the date of maximum brightness. Record the image size here 6.0 mm mm.

3. Using Figure 27-1, estimate the apparent magnitude (m) of SN1987 A at maximum brightness. At maximum brightness, SN1987 A had a magnitude of 1.

4. Using the formula M = m + 5 − 5 log r, calculate the absolute magnitude (M) of the supernova at peak brightness. Its absolute magnitude was −17.49.

Show calculations and answers to discussion questions here.

1 + 5 − 23.49

Name ______________________ Date ______________

Exercise 27. Supernova 1987 A

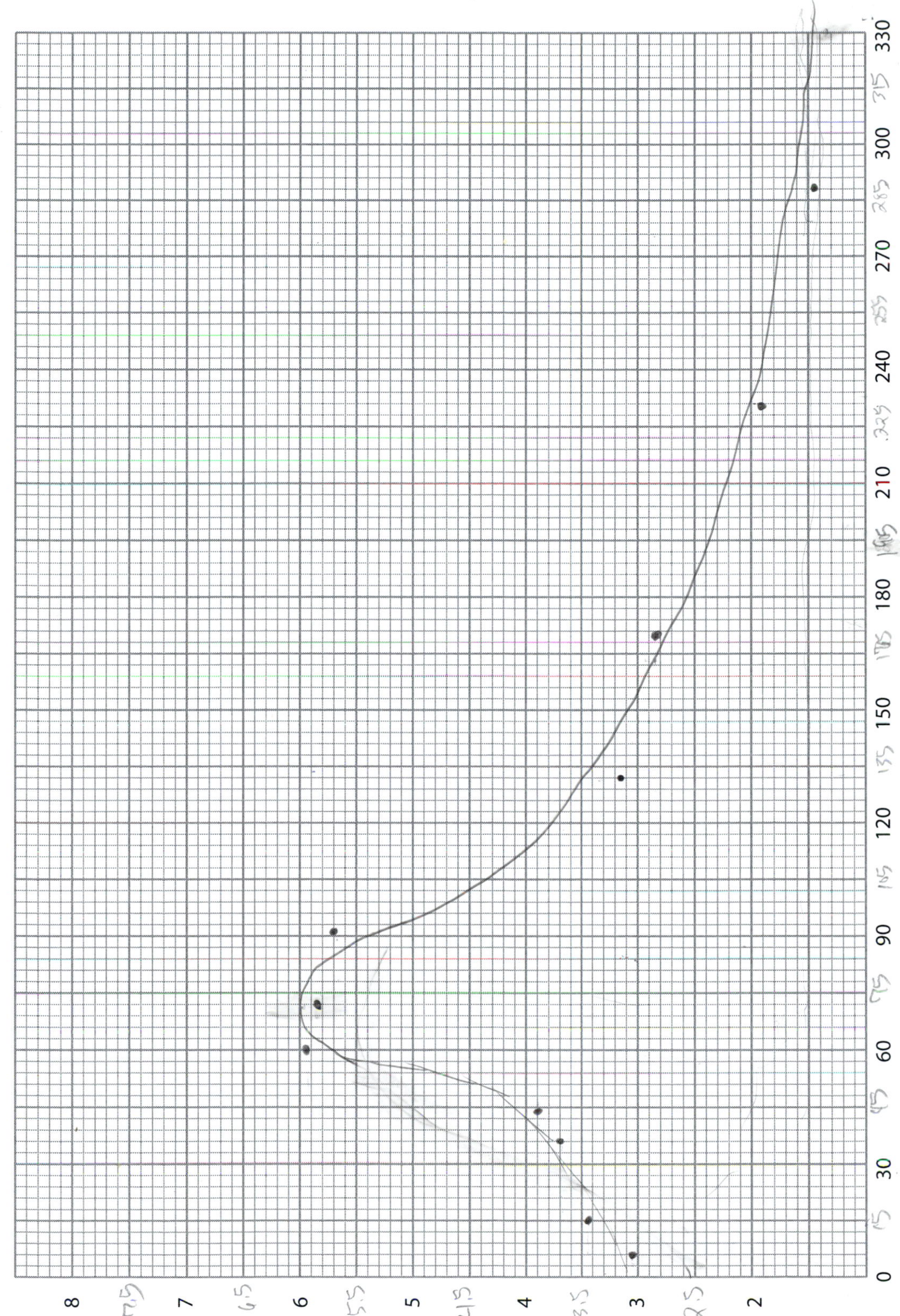

Discussion Questions

1. Based on your work, do you believe that SN 1987A was a Type I or Type II supernova?

2. What are some possible sources of error in this experiment?

3. About how long did it take for SN 1987A to reach maximum brightness from the point of first observation? About how long did it take to return to the brightness it had when first observed? What might be some reasons for one of these times to be longer than the other?

Exercise 28

Galactic Clusters and HR Diagrams[1,2]

Purpose and Processes

The purpose of this exercise is to study the relative HR diagrams (or color-magnitude diagrams) of several open clusters to determine their ages and distances, and to detect the effect of interstellar reddening on cluster observations. The processes stressed in this exercise include:

Questioning
Interpreting Data
Using Numbers
Formulating Models
Inferring
Using Logic
Predicting

Introduction

In the previous exercises on the Pleiades we have seen that a color-magnitude or HR diagram of a star cluster allows us to study stellar properties in general, and to determine cluster properties such as distance. The study of a number of galactic clusters also has produced some very important information about stellar evolution. The *zero age main sequence* (ZAMS) is the locus of points on the HR diagram where stars of different masses first begin to generate all their energy by the fusion of hydrogen.

The ZAMS therefore can be defined by the color-magnitude diagrams of some very young open clusters. This process also can be reversed to determine the ages and distances of many other galactic clusters. This method compares a cluster's color-magnitude diagram to the ZAMS on a standard HR diagram.

Figures 28-1 through 28-9 show the color-magnitude diagrams for nine open clusters. Note that the vertical axes give the apparent visual magnitude (V) of the stars in the clusters rather than the absolute magnitude (M_v) found in most HR diagrams. Color-magnitude diagrams are sometimes called "relative" HR diagrams because the apparent

[1]This exercise is adapted with permission from one developed by John D. Fix formerly at The University of Iowa.
[2]Cluster diagrams and ZAMS overlay are from Gretchen L. Hagen, Publications of the David Dunlap Observatory, *4* (1970); by permission of the author.

magnitude depends on distance as well as the intrinsic brightness of stars.

The top axis gives $(B-V)_O$, the intrinsic color index (indicating spectral type) of each star in the cluster. The bottom axis gives the color index (B-V) for each star as actually measured. The reason for the difference in these scales will be examined later in this exercise.

Figure 28-10 (in Appendix 4) is a translucent overlay showing the ZAMS on an HR diagram. Note that here the axes are labeled absolute magnitude and intrinsic color.

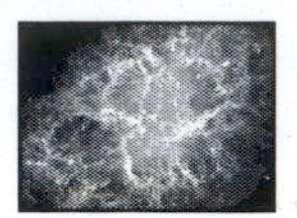

Procedure

CLUSTER AGE

We can assume that all stars in a given cluster have about the same age and original chemical composition. All stars of a cluster don't fall on the main sequence, however, because stars of different masses evolve at different rates. In addition to main sequence stars, a cluster may contain stars that are just evolving away from the main sequence, such as red giants. In general, we might expect that in an older cluster more stars would have had time to evolve from the main sequence. From studies of stellar evolution we find that this is just the case: the more massive O- and B-type stars leave the main sequence first. By finding the *turn-off point,* that region of the main sequence where the stars are just beginning to evolve toward the red giant region, we can estimate the cluster's age. Figure 28-11 shows a graph of $(B-V)_O$ of this turn-off point versus the age for clusters.

1. Place the translucent overlay over the HR diagram of each cluster and slide it left or right until the $(B-V)_O$ scales for the two diagrams match. Now slide the overlay up and down (keeping the $(B-V)_O$ scales aligned) until you get the best possible match between the ZAMS and the main sequence for each cluster. Remembering that hot main sequence stars (having small or negative color indices) leave the main sequence first, be sure that the cooler main sequence stars fall on the ZAMS even if the hotter stars do not.
2. After matching the ZAMS and the cluster main sequence, find the turn-off point of the main sequence for the cluster. This point should be carefully chosen so that there are few stars lying to the left of it. Use Figure 28-11 to determine the age of each cluster.

CLUSTER DISTANCE

The M_V scale of the HR diagram overlay and a cluster's V scale will match only if the cluster is ten parsecs away because a star's absolute magnitude is the apparent magnitude it would have if it were at a distance of ten parsecs. In fact, none of these clusters is as close as ten parsecs, so the apparent magnitudes are larger (indicating the stars are dimmer) than the corresponding absolute magnitudes on the ZAMS overlay. By comparing the corresponding V and M_V we can determine cluster distances in parsecs using the relationship

$$\log d = \frac{V - M_V + 5}{5}$$

where d = cluster distance in pc
V = apparent visual magnitude
M_V = absolute visual magnitude.

Matching the cluster and ZAMS $(B\text{-}V)_O$ scales as in *Cluster Age,* read the corresponding values of M_V and V. Find $V\text{-}M_V$ (the *distance modulus*) for each cluster and determine the distance to each using either the equation above or the graph of distance modulus vs. distance in Figure 28-12.

COLOR EXCESS AND REDDENING

As noted earlier, the (B-V) and $(B\text{-}V)_O$ scales of the cluster diagrams do not line up: in all cases the lower (B-V) scale is shifted to the left of the upper $(B\text{-}V)_O$ scale. For example, we might measure the (B-V) for a star to be 0.4, yet on the upper scale we see its intrinsic color index is 0.3. We are observing the stars in a cluster to be redder than they really are because larger positive numbers refer to redder stars.

This difference between apparent or measured color index and the intrinsic color index is called the *color excess* of a cluster and can be written

$$CE_{B\text{-}V} = (B\text{-}V) - (B\text{-}V)_O.$$

1. Measure the color excess for each of the nine clusters.
2. Table 28-1 gives the actual distances to the nine clusters. How do your distance determinations compare to the real distances? How can you account for the discrepancies (if any) that exist?
3. Calculate the ratio of the distance you found to the actual distance of each cluster. Make a graph showing this ratio versus $CE_{B\text{-}V}$. Is this ratio related to the color excess of a cluster? If so, how do you account for this fact?

Table 28-1. Cluster Distance

Cluster		Distance
NGC	457	2900 parsecs
NGC	752	360
Mel	20	170
M	45	120
NGC	2632	160
NGC	2682	800
IC	4725	600
NGC	6705	1700
NGC	6791	5200

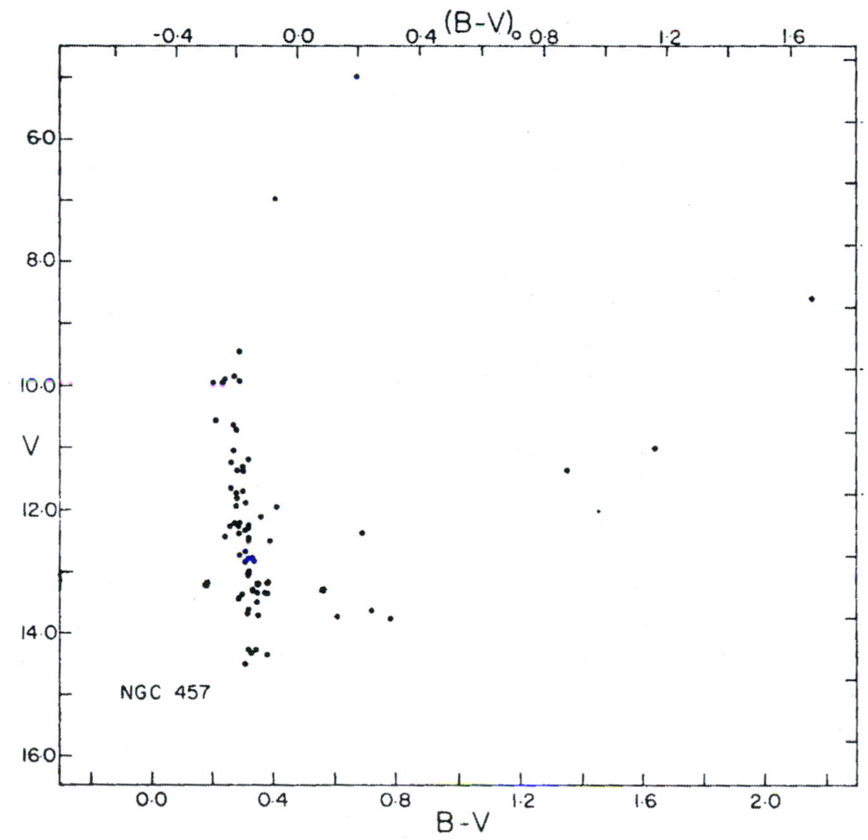

Figure 28-1. A color-magnitude diagram of NGC 457. (From *An Atlas of Open Cluster Colour-Magnitude Diagrams* by Gretchen L.H. Harris. Copyright © 1970 by Gretchen L.H. Harris. Reprinted by permission.)

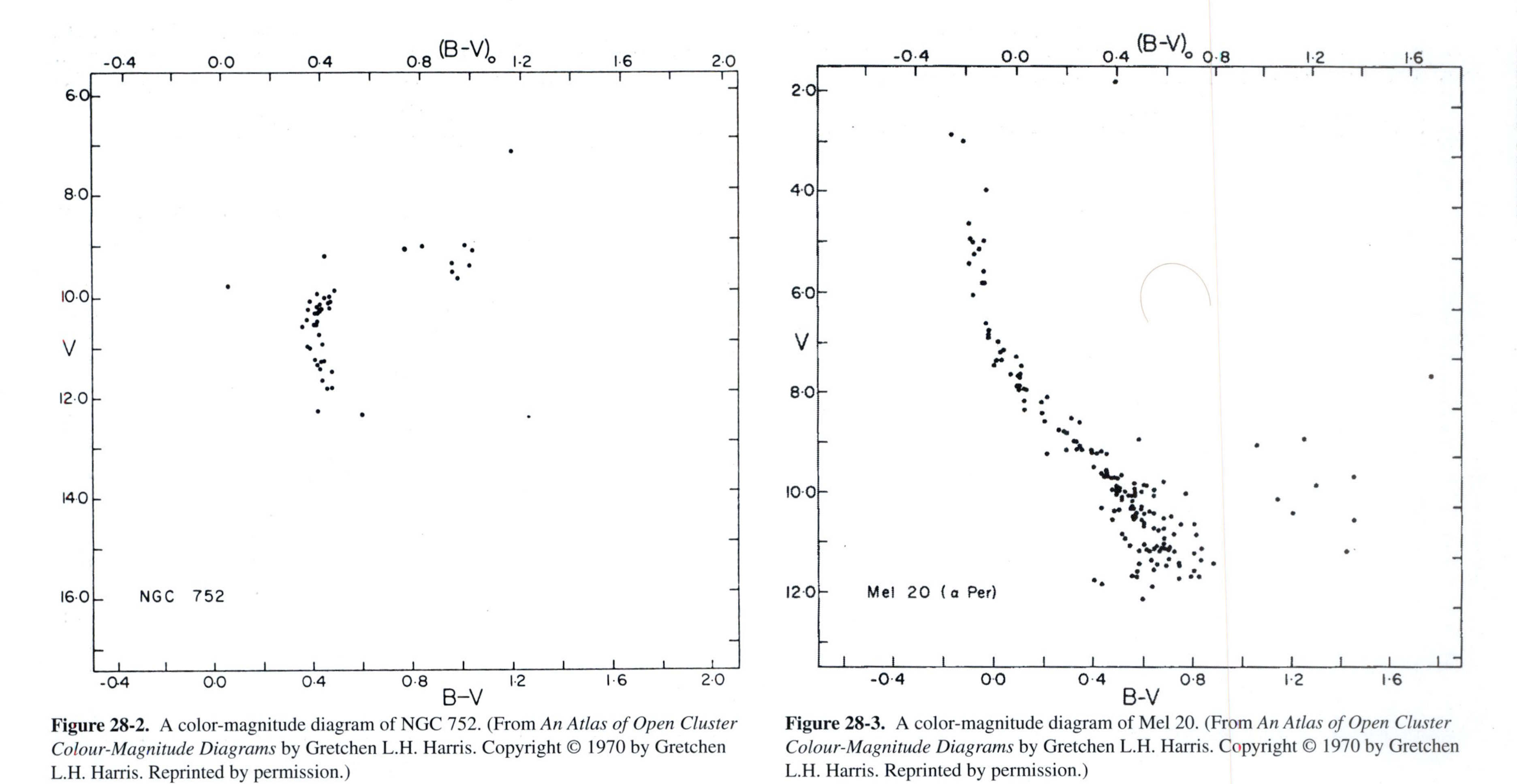

Figure 28-2. A color-magnitude diagram of NGC 752. (From *An Atlas of Open Cluster Colour-Magnitude Diagrams* by Gretchen L.H. Harris. Copyright © 1970 by Gretchen L.H. Harris. Reprinted by permission.)

Figure 28-3. A color-magnitude diagram of Mel 20. (From *An Atlas of Open Cluster Colour-Magnitude Diagrams* by Gretchen L.H. Harris. Copyright © 1970 by Gretchen L.H. Harris. Reprinted by permission.)

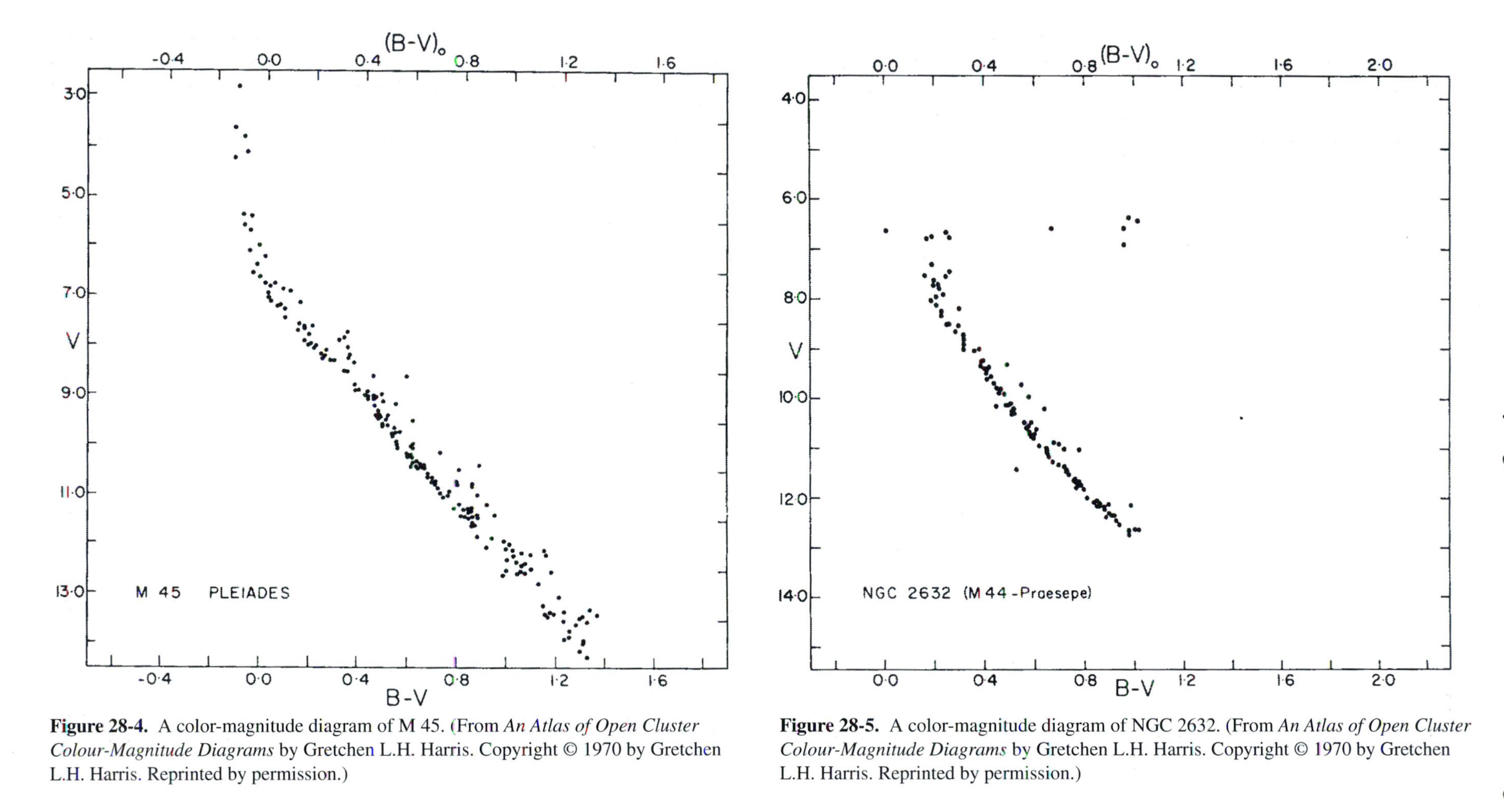

Figure 28-4. A color-magnitude diagram of M 45. (From *An Atlas of Open Cluster Colour-Magnitude Diagrams* by Gretchen L.H. Harris. Copyright © 1970 by Gretchen L.H. Harris. Reprinted by permission.)

Figure 28-5. A color-magnitude diagram of NGC 2632. (From *An Atlas of Open Cluster Colour-Magnitude Diagrams* by Gretchen L.H. Harris. Copyright © 1970 by Gretchen L.H. Harris. Reprinted by permission.)

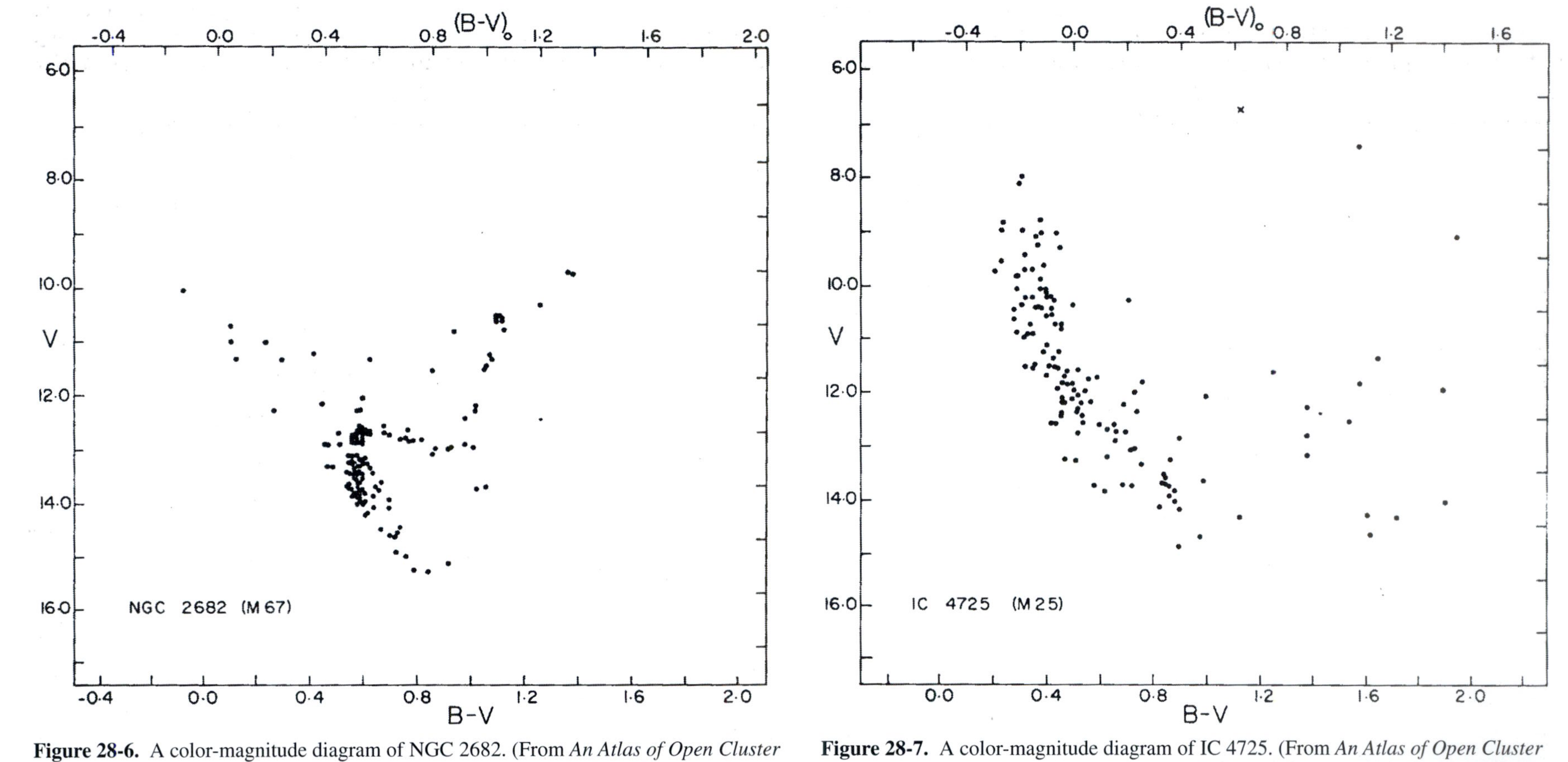

Figure 28-6. A color-magnitude diagram of NGC 2682. (From *An Atlas of Open Cluster Colour-Magnitude Diagrams* by Gretchen L.H. Harris. Copyright © 1970 by Gretchen L.H. Harris. Reprinted by permission.)

Figure 28-7. A color-magnitude diagram of IC 4725. (From *An Atlas of Open Cluster Colour-Magnitude Diagrams* by Gretchen L.H. Harris. Copyright © 1970 by Gretchen L.H. Harris. Reprinted by permission.)

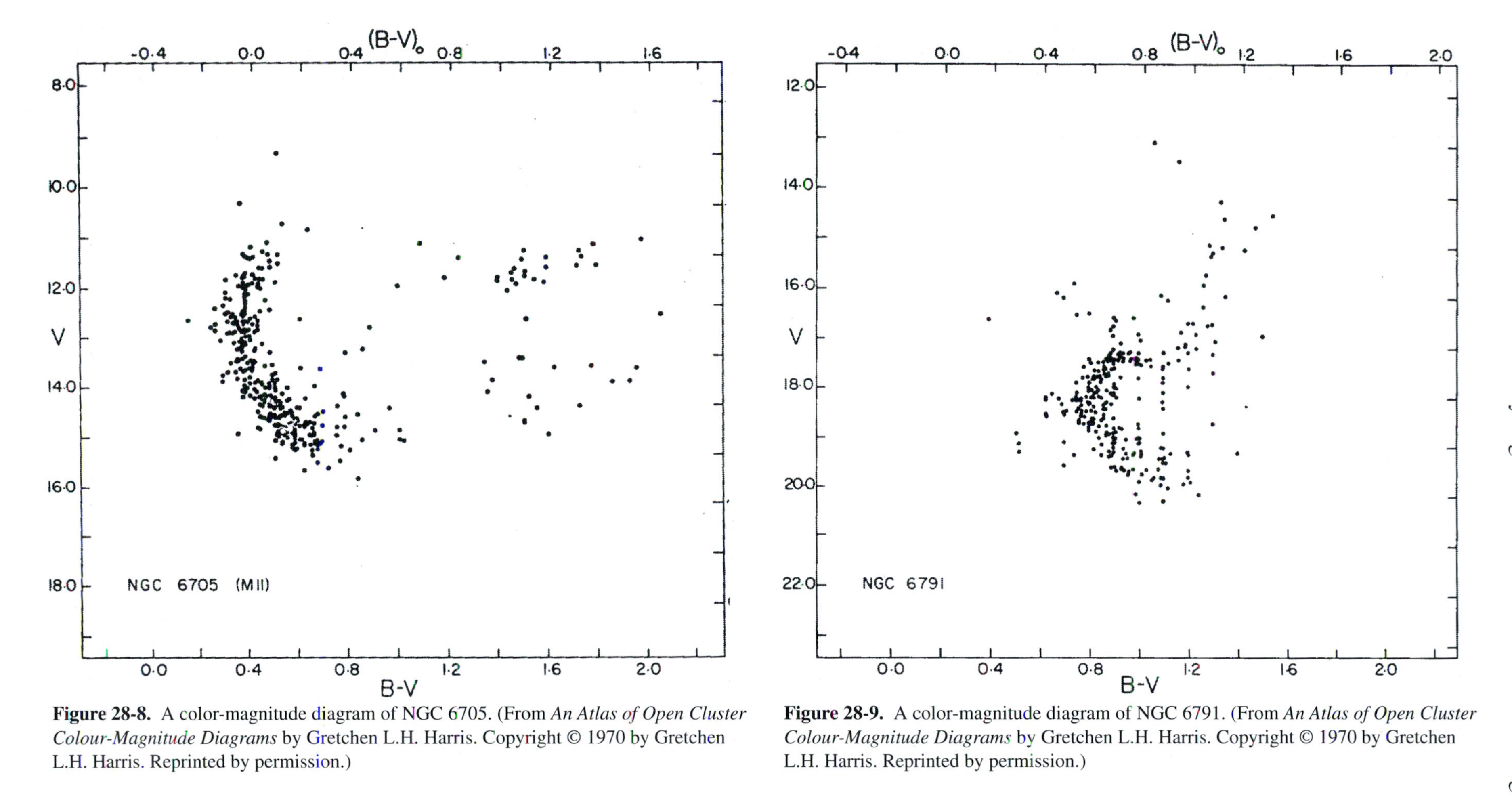

Figure 28-8. A color-magnitude diagram of NGC 6705. (From *An Atlas of Open Cluster Colour-Magnitude Diagrams* by Gretchen L.H. Harris. Copyright © 1970 by Gretchen L.H. Harris. Reprinted by permission.)

Figure 28-9. A color-magnitude diagram of NGC 6791. (From *An Atlas of Open Cluster Colour-Magnitude Diagrams* by Gretchen L.H. Harris. Copyright © 1970 by Gretchen L.H. Harris. Reprinted by permission.)

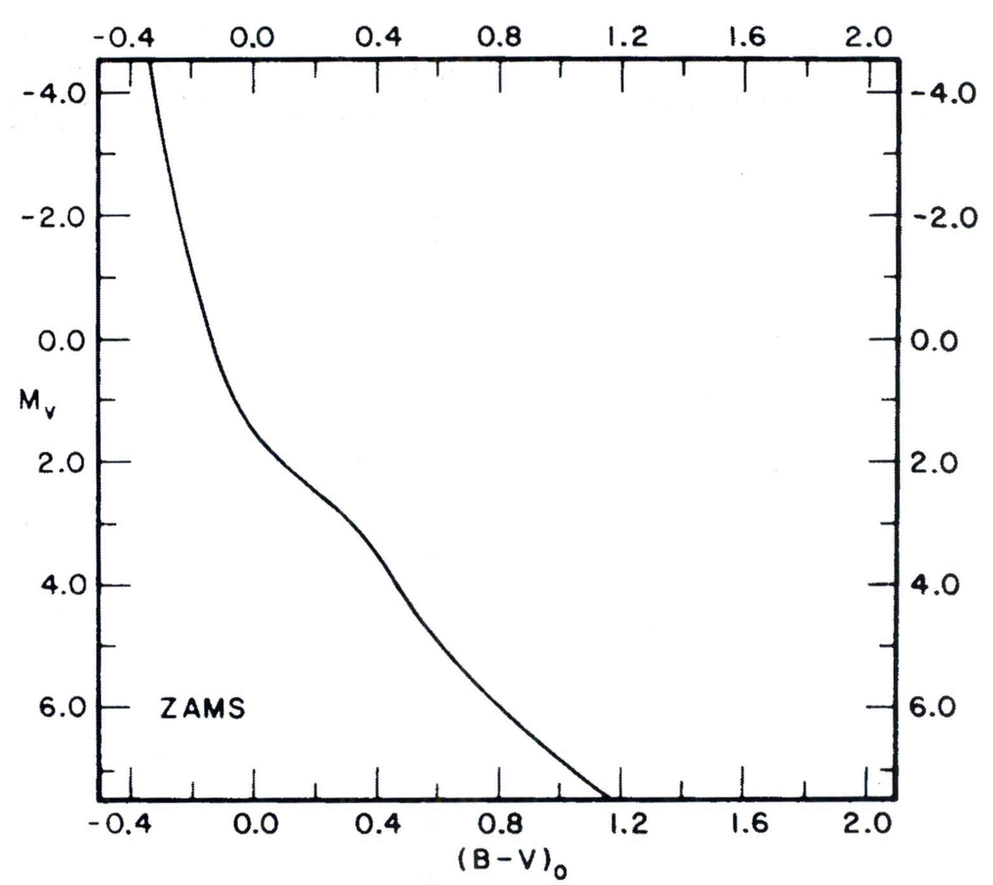

Figure 28-10. The zero age main sequence. (From *An Atlas of Open Cluster Colour-Magnitude Diagrams* by Gretchen L.H. Harris. Copyright © 1970 by Gretchen L.H. Harris. Reprinted by permission.)

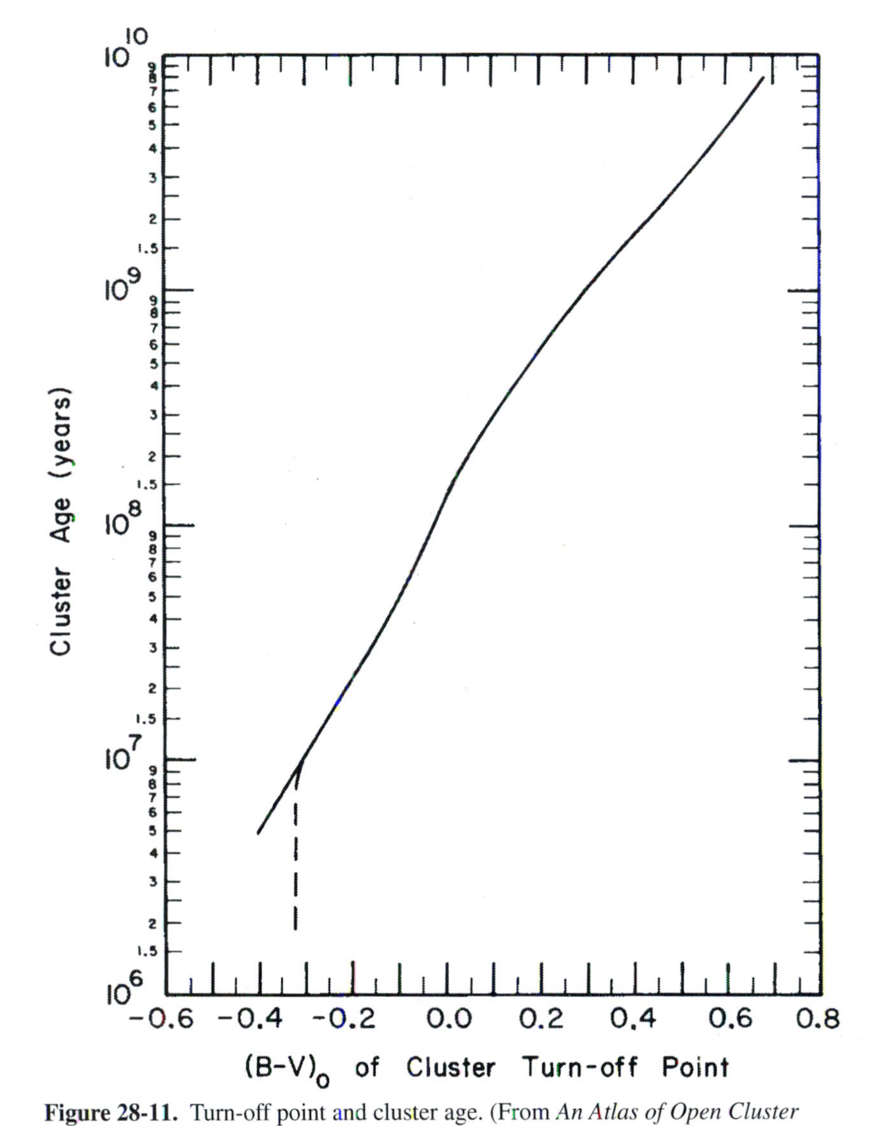

Figure 28-11. Turn-off point and cluster age. (From *An Atlas of Open Cluster Colour-Magnitude Diagrams* by Gretchen L.H. Harris. Copyright © 1970 by Gretchen L.H. Harris. Reprinted by permission.)

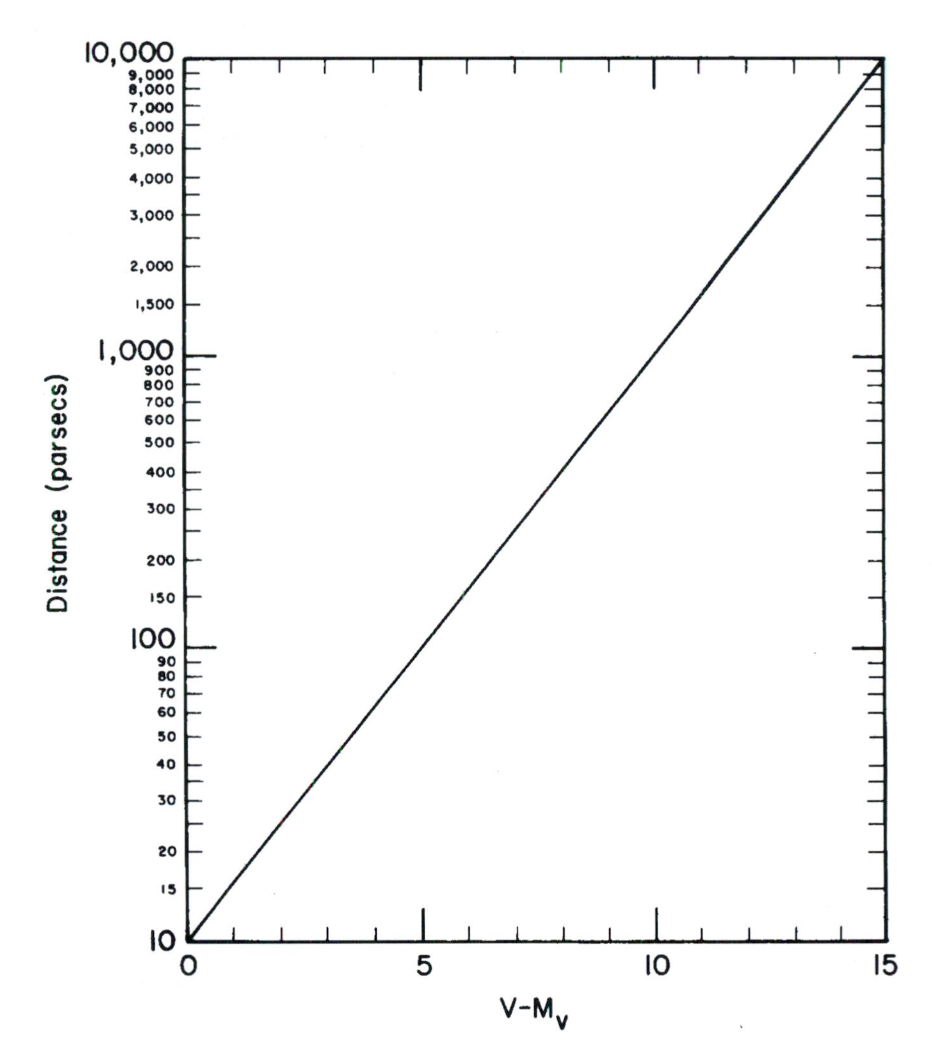

Figure 28-12. Distance modulus and distance. (From *An Atlas of Open Cluster Colour-Magnitude Diagrams* by Gretchen L.H. Harris. Copyright © 1970 by Gretchen L.H. Harris. Reprinted by permission.)

Discussion Questions

1. Would you expect to see many clusters younger than 10^7 years or older than 5×10^9 years? Why or why not?

2. In this exercise you were given $(B\text{-}V)_0$ for each cluster. How might this value be determined?

3. All globular clusters have HR diagrams that are qualitatively similar to the HR diagram of the galactic cluster M 67, shown in Figure 28-6. Can you use this information to deduce anything about globular clusters?

4. Recap what you found regarding the relationship between color excess and distance over-estimation. Given that you overestimated the cluster distances, what techniques might be used to determine the accurate distance?

Exercise 29

The Distribution of Star Clusters on the Sky

Purpose and Processes

The purpose of this exercise is to use the observed spatial distribution of star clusters to deduce key features of the Milky Way's geometry.

Using Numbers
Interpreting Data
Using Logic
Inferring
Formulating Models

Introduction

Astronomers often study the distributions of objects on the sky. That is, they will study a specific class of objects to see what can be learned from where we see them. One might learn something about the objects being studied or about the larger system containing the objects. For example, in Exercise 3 you were asked to plot the positions of the solar system's planets on the star charts found at the back of this book. The first discussion question of that exercise asked what one might deduce about the geometry of the solar system from the pattern the planets made against the background stars. The apparent distribution of the planets yielded little, if any, information about the physical nature of the planets but was quite useful for learning about the geometry of the solar system in which the planets reside.

In a relatively recent example, astronomers attempted to learn about the nature of gamma ray bursts (GRBs) by studying the distribution of these events on the sky. Discovered in 1970, GRBs appear as intense flashes of gamma rays that typically last a few seconds. Most early models for the physical nature of the objects causing these flashes involved neutron stars. The neutron stars in our galaxy, however, lie in a distribution that is strongly aligned with the Milky Way's disk. As the positions of more and more GRBs were determined it became apparent that their distribution was inconsistent with the distribution of known neutron stars. The positions of these GRBs appeared to be consistent with isotropy. When a distribution of objects is isotropic, the objects are equally likely to be seen in any direction one looks. Astronomers theorized that these objects could be associated with a large spherical halo around the galaxy since any other distribution within the galaxy would not appear isotropic to us due to our location within the Milky

Way. Eventually measurements unrelated to distribution studies showed that the GRBs were associated with extremely distant galaxies. The observed GRB distribution arose because the large-scale distribution of galaxies is isotropic. It should be observed that the apparent isotropy of the GRBs led some astronomers (notably Bohdan Paczynski) to postulate extreme distances for the GRB sources early on.

The previous examples highlight the importance of distribution studies in astronomy. It should also be apparent that these studies must be done carefully since they are prone to observational biases and a large statistical sample is frequently required to make any definitive claims.

In this exercise you will map the distribution of bright star clusters to look for patterns that could tell us something about the physical nature of the star clusters, the location of the clusters or the nature of the system containing the clusters. The clusters we will use are all part of the famous list Charles Messier compiled while searching for comets from about 1757 to about 1783 in France. We divide the clusters into two groups: globular clusters and open (also know as galactic) clusters. An example of each type of cluster is shown in Figure 30-1. Globular clusters have a much higher star density and greater total number of stars than open clusters.

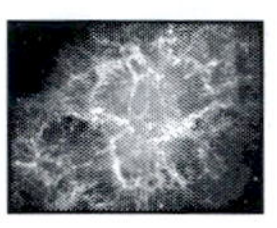

Procedure

1. The right ascension and declination of each open star cluster on Messier's list can be found in Table 29-1. Use these measurements to create a plot of declination versus right ascension for these clusters on the graph titled *Messier Open Clusters.*
2. The right ascension and declination of each globular star cluster on Messier's list can be found in Table 29-2. Use these measurements to create a plot of declination versus right ascension for these clusters on the graph titled *Messier Globular Clusters.*
3. Creating a histogram is a useful way to search for clumping in any distribution. Such clumping will appear as a peak in the histogram. Make a histogram by plotting the number of globular clusters in each hour of right ascension on the graph titled *Messier Globular Cluster Right Ascensions.* A histogram would typically be made by drawing a line segment across the hour of right ascension at the appropriate level for the number of clusters with that right ascension. Then vertical line segments connect these horizontal segments at the edges of each right ascension bin.
4. From photographs such as those in Figure 29-1 it is difficult to be certain whether or not globular clusters and open clusters are different populations of objects. It is possible that globular clusters are merely very rich examples and open clusters sparse examples of the same type of cluster. Measurements beyond simple photographs must be made. Your distribution maps are one additional measure. Would you say that your maps support a model of globular clusters and open clusters as distinct classes of objects or a model of globular clusters and open clusters as examples of the same type of object? Does it appear that either distribution is isotropic? What might that tell you about where these objects are? Explain your answers to all these questions.

Table 29-1. Open Cluster Coordinates

Cluster	R.A.(hrs)	Dec.(deg)	Cluster	R.A.(hrs)	Dec.(deg)
M6	17.663	−32.217	M37	5.8680	32.550
M7	17.893	−34.817	M38	5.4750	35.833
M8	18.058	−24.383	M39	21.533	48.417
M11	18.848	−6.2670	M41	6.7630	−20.733
M16	18.308	−13.783	M44	8.6650	19.980
M18	18.328	−17.133	M45	3.7900	24.117
M21	18.073	−22.500	M46	7.6930	−14.800
M23	17.942	−19.017	M47	7.6080	−14.483
M24	18.282	−18.483	M48	8.2270	−5.7830
M25	18.527	−19.2500	M50	7.0500	−8.3333
M26	18.750	−9.4000	M52	23.4000	61.567
M29	20.395	38.517	M67	8.8700	11.833
M34	2.6950	42.767	M73	20.978	−12.617
M35	6.1430	24.333	M93	7.7420	−23.850
M36	5.5980	34.133	M103	1.5500	60.683

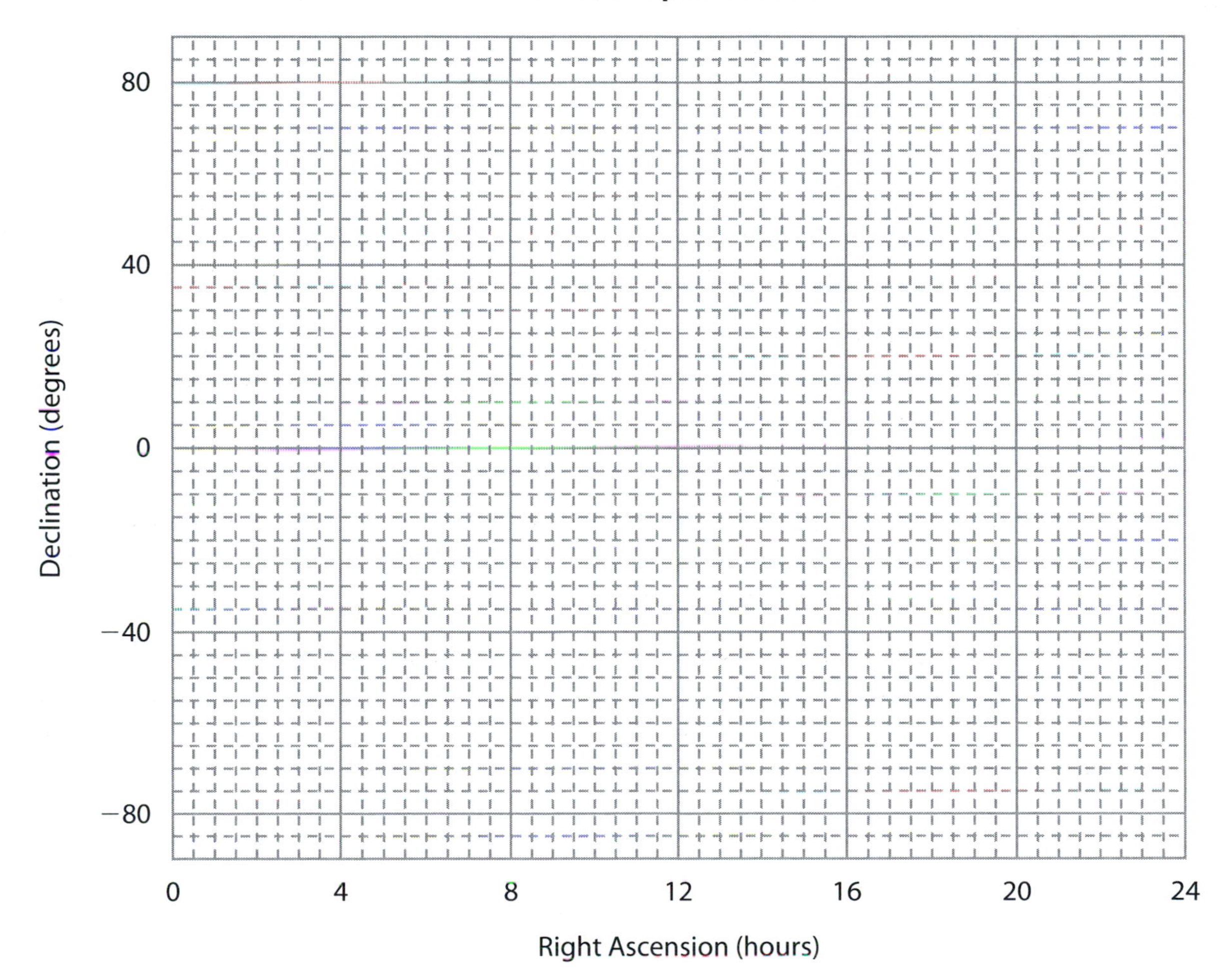

Table 29-2. Globular Cluster Coordinates

Cluster	R.A.(hrs)	Dec.(deg)	Cluster	R.A.(hrs)	Dec.(deg)
M2	21.555	−0.83300	M54	18.932	−30.483
M3	13.700	28.400	M55	19.662	−30.983
M4	16.388	−26.520	M56	19.275	30.017
M5	15.307	2.0830	M62	17.015	−30.117
M9	17.148	−18.517	M68	12.653	−26.733
M10	16.948	−4.1000	M69	18.518	−32.350
M13	16.692	36.467	M70	18.717	−32.300
M14	17.623	−3.2500	M71	19.893	18.767
M15	21.497	12.150	M72	20.888	−12.550
M19	17.043	−26.250	M75	20.098	−21.917
M22	18.602	−22.900	M79	5.4050	−24.550
M28	18.403	−24.867	M80	16.278	−22.967
M30	21.670	−23.200	M92	17.283	43.133
M53	13.212	18.017	M12	16.783	−1.9500

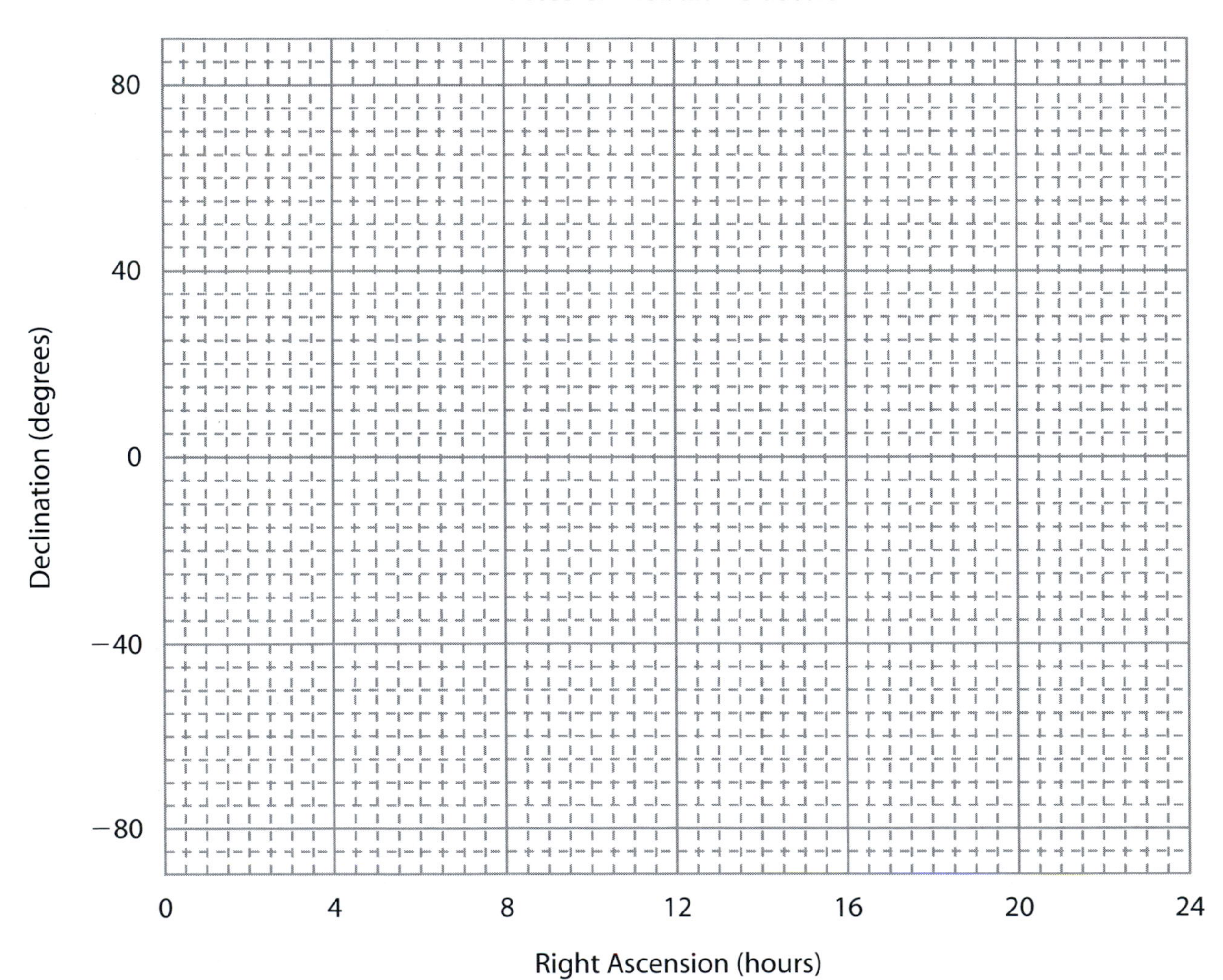

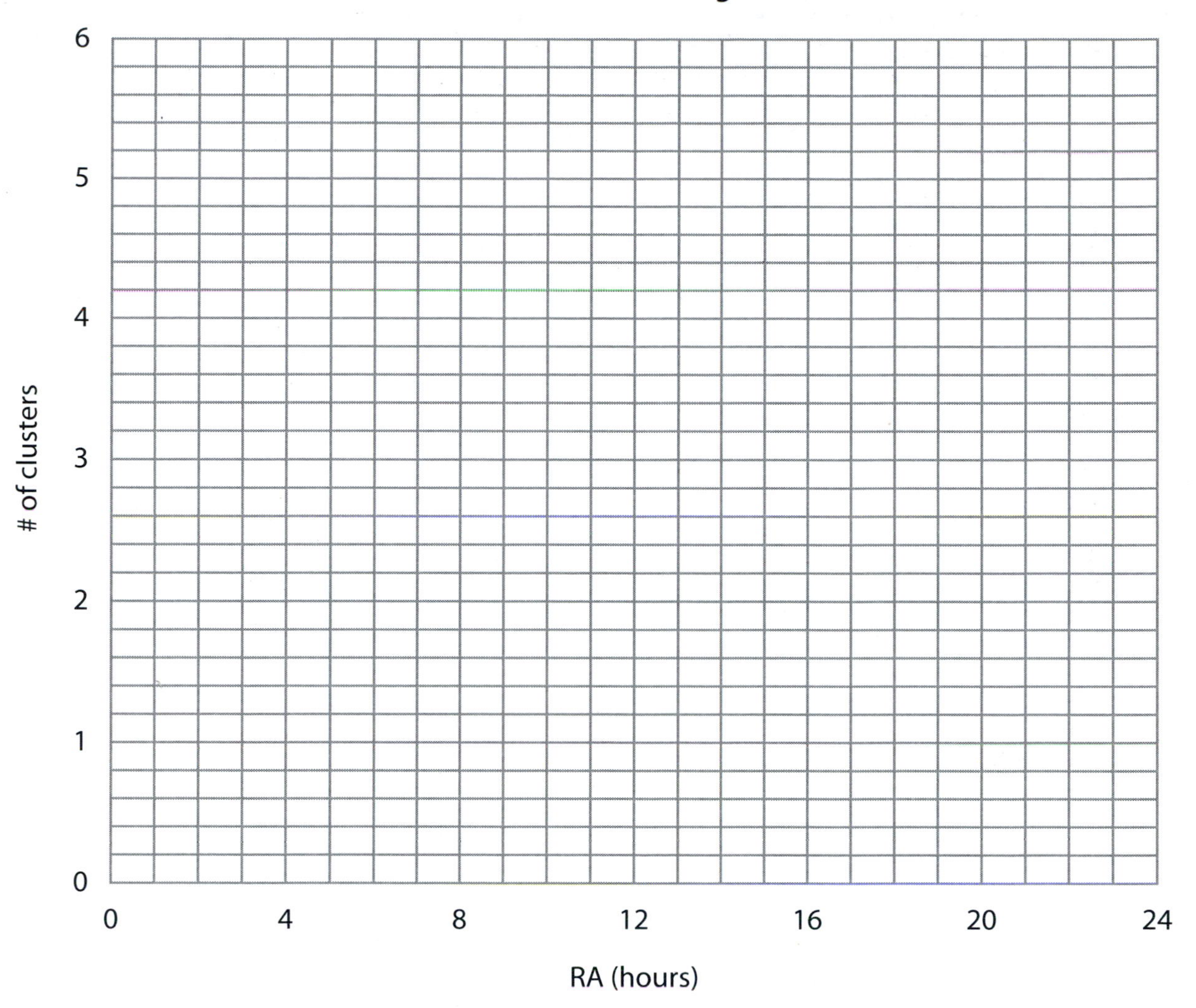
Messier Globular Cluster Right Ascensions
# of clusters
0
1
2
3
4
5
6
RA (hours)
0
4
8
12
16
20
24

Figure 29-1a. Globular cluster M4.

Figure 29-1b. Open cluster M23.

5. In 1917 Harlow Shapley, working at Harvard, set about determining the size of the Milky Way galaxy. His first step in this difficult undertaking was to hypothesize that globular clusters were distributed in a clump around he center of the galaxy. If this hypothesis is true, you can estimate the position of the center of the galaxy on the sky. It should simply be the middle of the distribution of globular clusters. Determine the mean right ascension and mean declination for the clusters in Table 29-2. Compare your results to the currently accepted position of the galactic center on the sky of right ascension = 17h45m and declination = −29°00′. If the globular clusters are truly centered on the galactic center what might account for any discrepancies between the position you determined and the actual position of the galactic center?

Discussion Questions

1. What might be the cause of any gaps that appear in your graph of right ascension versus declination for the open clusters?

2. What can you deduce about where open clusters exist based on the pattern of their distribution as shown by the graph of right ascension versus declination?

3. At least one cluster appears to be an "outlier" in the globular cluster distribution, having a right ascension well away from the main peak. Can you use this observation to deduce anything special about where in space this cluster likely is?

4. Assuming that globular clusters are in a spherical halo about the center of the galaxy, is the orbit of the sun well inside or mostly outside the edge of the distribution? Explain.

5. Suppose you wish to design an experiment to look for dark objects in the outer solar system as they pass in front of distant stars. To be successful you might want to look at open star clusters (because they have many bright stars) that lie in the plane of the earth's orbit. Use the star charts found in the back of this book to suggest a few constellations that might be the good places to focus your search.

6. There are a great many clusters in the sky that are not on Messier's list. Explain how your results in this activity would be altered as you added more and more of these clusters to your study.

Exercise 30

Galaxies in the Virgo Cluster

Purpose and Processes

The purpose of this exercise is to develop a system for classifying galaxies, to compare the numbers of different types of galaxies in the Virgo cluster, and to determine the linear sizes of some of the cluster members. The processes stressed in this exercise include:

Classifying
Using Numbers
Predicting

Introduction

The Virgo cluster of galaxies is the nearest of the large clusters and is probably one of the largest. It covers an area of about ten degrees by twelve degrees in the sky and is centered near the Virgo-Coma constellation border. Nearly 3,000 members of the cluster have been identified on high-resolution photographs, and several are bright enough to be listed in the Messier catalogue.

The advantages of studying a cluster of galaxies are similar to those in studying star clusters: the objects are all at about the same distance, so relative luminosities and diameters can be compared for the members. As with stellar evolution, in studying galaxies sharing a common origin we hope to gain clues about the processes that govern the formation and evolution of different types of galaxies. We can also see how our galaxy or local group of galaxies compares with others in space.

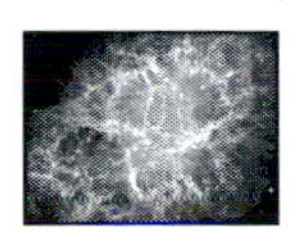

Procedure

Figures 30-1 through 30-4 are reprinted from a Palomar Sky Survey plate showing the Virgo cluster of galaxies. Other photos or slides of galaxies will be available in the classroom.

CLASSIFICATION OF GALAXIES

Examine as many photographs of galaxies as you can and look for similarities and differences. Consider properties such as overall shape, prominence of the central bulge relative to the disc, and possible arm structure.

1. Devise a classification system for different types of galaxies. Use major classes and subclasses if they seem appropriate.
2. Devise a system of names or numbers to identify different classes.
3. Could any of your classes represent the objects in other classes seen from different angles? That is, could one class be the same objects as another class but seen from a different point of view? Examine your classes for possible overlap due to different spatial orientations of the objects and revise your system, if necessary, to represent fundamentally different shapes of galaxies.

RELATIVE NUMBERS OF DIFFERENT TYPES OF GALAXIES

A system is needed for describing general plate locations to avoid counting galaxies more than once. One method is similar to that used for archeological field work: Divide the plate into a grid and label or number the squares.

1. Use your classification system to count and record the number of galaxies of each general class for the elements of your grid system. A hand magnifier may be helpful in distinguishing stars from giant galaxies. Galaxies have "fuzzy" edges while stars have clearer, more well-defined images.
2. Calculate the proportion or percent of each major type of galaxy in the cluster.

DISTANCE TO THE CLUSTER

Figure 30-5 shows the spectrum of one galaxy in the Virgo cluster with the major comparison lines of hydrogen and helium identified.

1. Determine the dispersion or plate scale for the galaxy spectrum in Å/mm. (You might want to refer to Exercise 19 *Evidence of the Earth's Revolution* for more detailed information on measuring Doppler shifts.)
2. Measure the displacement of the calcium H and K lines (indicated by the arrow) in millimeters and convert this shift to Angstroms.

3. Calculate the Doppler velocity for the spectrum using the relationship

$$v = c\frac{\Delta\lambda}{\lambda}$$

where v = velocity
c = speed of light = 3×10^5 km/sec
$\Delta\lambda$ = shift in wavelength in Å
λ = unshifted or rest wavelength.
= 3968 Å Ca II, H
= 3933 Å and K lines.

Calculate the distance to the Virgo cluster of galaxies using Hubble's Law. Recall that

$$r = \frac{V}{H_o}$$

where r = distance in megaparsecs
V = the recessional velocity in km/sec
H_o = Hubble's constant = 72 km/sec/Mpc.

SIZE OF THE GALAXIES IN THE CLUSTER

It would be interesting to compare the linear sizes of some of the Virgo cluster galaxies to that of our own galaxy. Therefore it is necessary to determine the angular size of the galaxies. Remember that

$$s = r\theta$$

where s = linear diameter
r = distance
θ = angular diameter in radians.

1. Measure the diameters of several of the galaxies to at least the 0.1 millimeter using a low power measuring microscope or an eyepiece and recticle.
2. The plate scale is 67.1″ of arc per mm.
3. Calculate the angular diameters of the galaxies measured above.
4. Calculate the linear diameters of the galaxies in kpc and compare these values to an estimated 30-kpc diameter for the Milky Way galaxy.

Figure 30-1. Portion of a Palomar Sky Survey plate of the Virgo cluster of galaxies. (Reprinted by permission of the Palomar Observatory.)

Figure 30-2. Portion of a Palomar Sky Survey plate of the Virgo cluster of galaxies. (Reprinted by permission of the Palomar Observatory.)

Figure 30-3. Portion of a Palomar Sky Survey plate of the Virgo cluster of galaxies. (Reprinted by permission of the Palomar Observatory.)

Figure 30-4. Portion of a Palomar Sky Survey plate of the Virgo cluster of galaxies. (Reprinted by permission of the Palomar Observatory.)

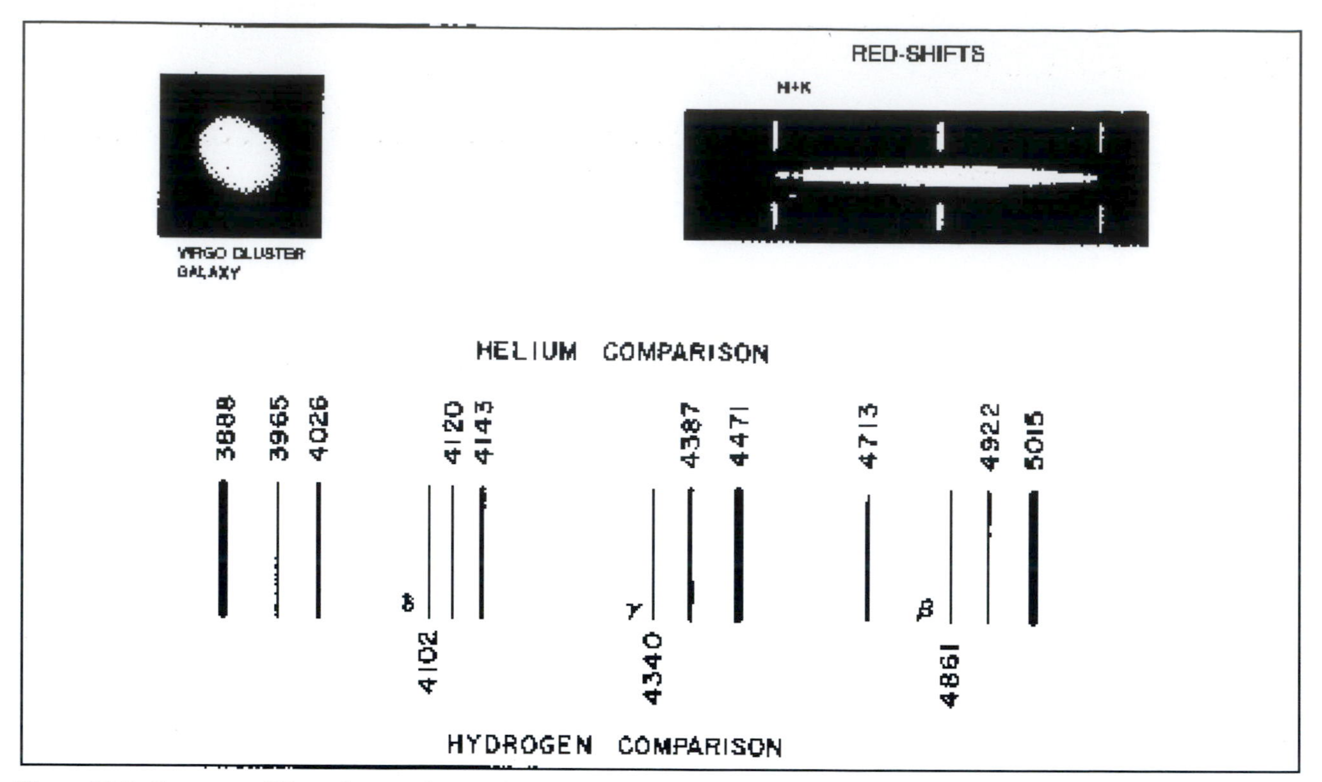

Figure 30-5. Spectrum of Virgo cluster galaxy and hydrogen comparison spectrum. (Reprinted by permission of the Palomar Observatory.)

Portrait of Margaret Geller

In the early 1980s, Dr. Margaret Geller and her colleague Dr. John Huchra, both of the Harvard-Smithsonian Center for Astrophysics, began the largest scale mapping of the universe. They obtained recessional velocities for a large number of galaxies and their corresponding precise locations. From the galaxies' recessional velocities they calculated the galactic distances out to a distance of about one billion light years. From this work they painstakingly produced the first large scale three-dimensional maps of the universe. The work is very slow and time-consuming and their first map was composed of a strip of the sky only 6 degrees wide and 153 degrees long. In this slice there were 1,099 clusters. They found that most galaxies surround voids that are roughly circular in cross section. This finding has led to a model of the universe that has the galaxies distributed in spherical thin sheets of space surrounding voids, and the name "bubble universe" has been coined to describe this distribution. (Photo by Darrel Hoff.)

Discussion Question

1. How would the ratios of different types of galaxies differ for a more distant cluster where only the brightest galaxies are seen?

Exercise

31

The Absolute Magnitude of a Quasar

Purpose and Processes

The purpose of this exercise is to show how a variety of astronomical techniques can be applied to a single problem in determining the absolute magnitude of a quasar. The processes stressed in this exercise include:

Using Numbers
Interpreting Data
Defining Operationally
Using Logic
Formulating Models
Predicting

Introduction

Stars produce a wide range of electromagnetic radiation. For example, the Sun (a typical star) radiates ultraviolet, visible, infrared, and radio waves. However, if the Sun were placed at stellar distances, we would not expect to be able to detect the comparatively feeble radio waves it produces. It was with great surprise, therefore, that two STELLAR radio sources were detected in 1960. These sources, unlike many other radio sources, could be positively associated with starlike images that had been previously photographed in the visible part of the spectrum. These sources were first named quasi-stellar radio sources, a name later shortened to *quasars.*

Quasar observations also were perplexing for a second reason: Their visible spectra showed emission lines that could not be identified with known chemical elements. In 1963 Martin Schmidt solved part of the problem by recognizing that the mysterious emission lines could be identified with known lines such as those of hydrogen if they were red shifted by very large amounts. However, if quasars obey accepted physical laws, such large red shifts would indicate that they must be receding at great velocities. Further, if they conform to Hubble's Law, all quasars must be at extremely great distances from our galaxy.

Following Schmidt's early work there was a great deal of controversy about the issue of quasars. If the objects were at cosmological distances they had to be intrinsically extremely bright objects. In the intervening years the issue has been resolved. It has been discovered that a number of quasars have been gravitationally lensed by closer (foreground) galaxies producing independent evidence for their great distances. In addition the Hubble Space Telescope has imaged the host galaxies for many quasars.

In this exercise we will investigate the properties of 3C273, one of the first two quasars discovered. We will assume it follows Hubble's Law and that its red shift is indicative of its distance. We will then arrive at an estimate of its luminosity and possible variability.

Procedure

Figure 31-1 shows the spectrum of quasar 3C273. The upper spectrum is that of the quasar and the lower spectrum is a comparison spectrum. The hydrogen Balmer lines (H_β, H_γ, and H_δ) are identified in the comparison spectrum and in their displaced positions in the quasar spectrum.

VELOCITY OF RECESSION

1. Determine the dispersion or plate scale for the quasar spectrum in Å/mm. (You might want to refer to Exercise 19 *Evidence of the Earth's Revolution* for more detailed information on measuring Doppler shifts.)

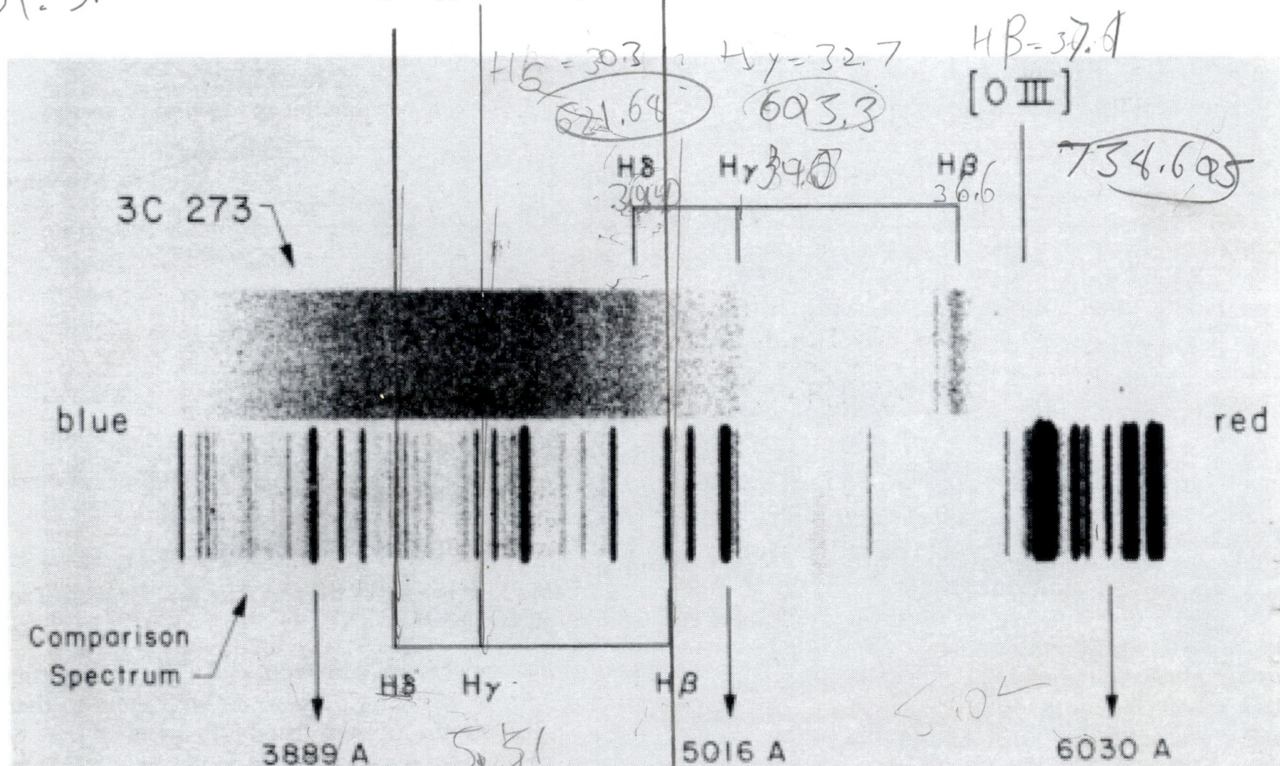

Figure 31-1. Spectrum of 3C273. (Reprinted by permission of the Palomar Observatory.)

2. Measure the displacement of one of the hydrogen lines in the spectrum and convert this shift to Angstroms.
3. Calculate the Doppler velocity for your spectral line. Remember that

$$v = c \frac{\Delta\lambda}{\lambda}$$

where v = velocity
c = speed of light = 3×10^5 km/sec
$\Delta\lambda$ = shift in wavelength in Å
λ = unshifted or rest wavelength in Å.

The rest wavelengths for the hydrogen Balmer lines are:

H_α = 6563 Å H_γ = 4340 Å
H_β = 4861 Å H_δ = 4102 Å

Repeat your measurements and calculations for the other hydrogen lines and average your results.

Note that for objects traveling at very high velocities this conventional Doppler equation must be replaced by the relativistic form. However, 3C273 is traveling slowly enough that the difference is not significant.

DISTANCE

Calculate the distance to 3C273 using Hubble's Law. Recall that

$$r = \frac{V}{H_o}$$

where r = the distance in megaparsecs
V = the recessional velocity in km/sec
H_o = Hubble's constant = 72 km/sec/Mpc.

ABSOLUTE MAGNITUDE AND ENERGY OUTPUT

1. Apparent Magnitude

Figure 31-2 shows a portion of the Palomar Sky Survey plate which contains 3C273 (marked with a black cross). This plate is a negative print on which bright objects appear dark and vice versa. Other stars in the field are marked with letters of the alphabet. Their apparent magnitudes are given in Table 31-1.

Estimate the apparent magnitude of 3C273, remembering that apparent photographic magnitudes are roughly inversely proportional to image size. (A graph of image diameter as a function of apparent magnitude may be useful).

2. Absolute Magnitude

It is possible to calculate an object's absolute magnitude knowing its apparent magnitude and distance using the relationship

$$M = m + 5 - 5 \log r.$$

Using your values for r (in parsecs) and your estimated apparent magnitude, calculate the absolute magnitude of 3C273.

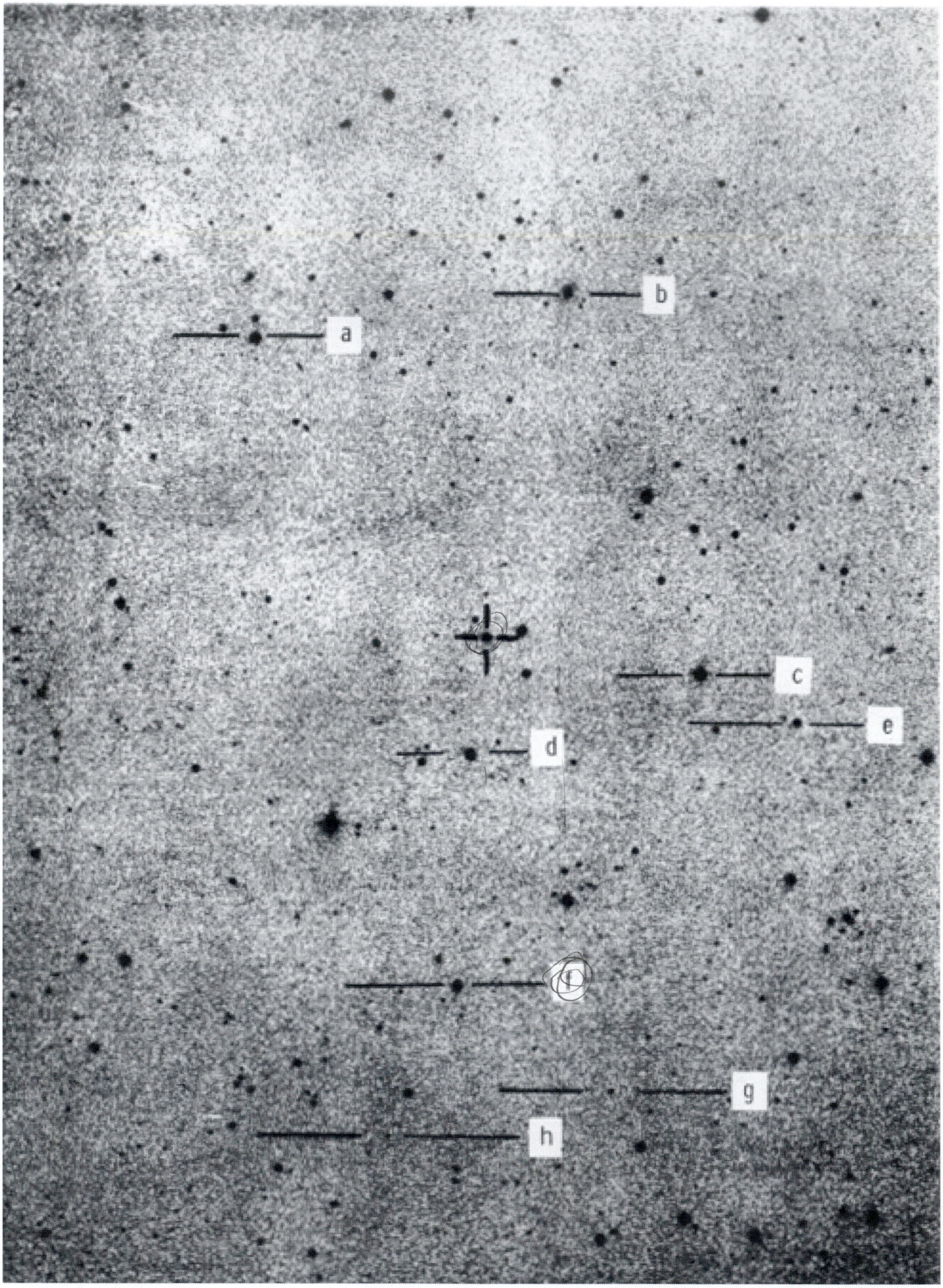

Figure 31-2. Photograph of 3C273 and the surrounding area of sky. (Reprinted by permission of the Palomar Observatory.)

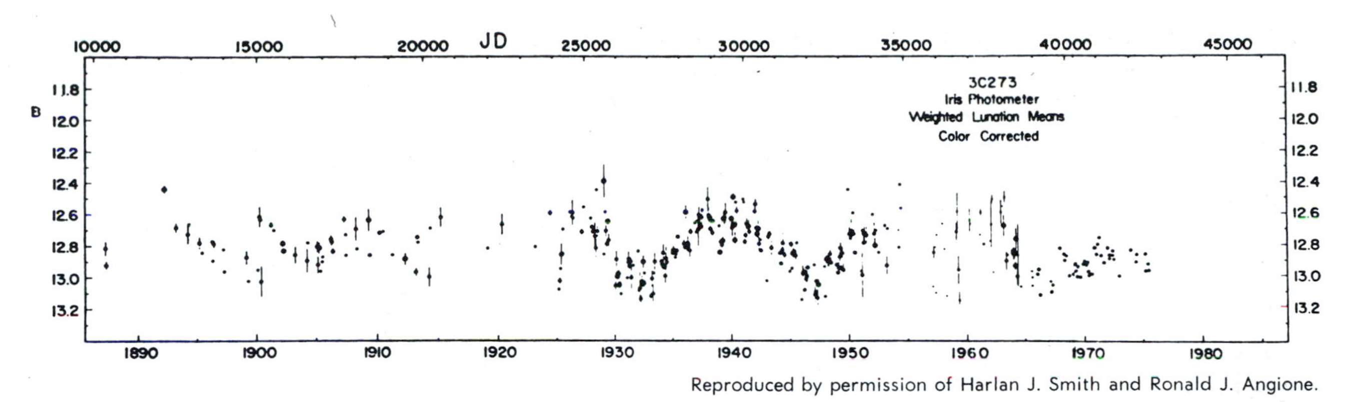

Reproduced by permission of Harlan J. Smith and Ronald J. Angione.

Figure 31-3. Brightness variations of 3C273.

Table 31-1. Comparison of Star Magnitudes

Star	Apparent Magnitude
a	12.5
b	13.1
c	12.1
d	12.6
e	13.2
f	12.8
g	13.7
h	14.0

Table 31-2. Comparison of Object Magnitudes

Object	Apparent Magnitude
Sun	+4.8
Sirius	+1.5
Canopus	−5.0
Deneb	−7.0
Globular Cluster	−9.0
Irregular Galaxy	−18.0
Spiral Galaxy	−21.0
Elliptical Galaxy	−23.0

3. Energy Output

Table 31-2 is a list of the absolute magnitudes of some familiar objects.

What conclusions can you make about the energy output of 3C273? Compare its brightness to that of several of the objects listed. (Remember that each magnitude difference corresponds to a brightness ratio of 2.5.)

VARIABILITY

Figure 31-3 shows the variation in apparent magnitude of 3C273 as measured from photographic plates since the late 1800s. The median value for these data corresponds with your measured apparent magnitude. Note the apparent regular variation in light output. It is generally accepted that the linear size of an object cannot exceed its period of variation. For example, an object with a period of 2 years can be no larger than 2 light years across.

1. Estimate the period of variation of 3C273. What is its maximum size?
2. Compare the absolute magnitude and size of 3C273 to those of other bright objects. Comment on this quasar's *energy density,* a quantity relating its energy output and size.

Portrait of Martin Schmidt

Dr. Martin Schmidt is a Dutch born (1929–) astrophysicist who shares with Dr. Jesse Greenstein the credit for unlocking the mystery of quasar distances. He dates his interest in the sky to a summer experience when he was 12 and an uncle let him look at the night sky through a small telescope. Later he finished his doctorate and joined the California Technical Institute staff in 1959. He was soon recognized as a diligent and skilled observer, and in 1962 he began observing quasars. At the time, the objects were regarded as a mystery because they appeared starlike, yet their spectra were not like those of other stars. In February, 1963, while working on a manuscript about the quasar 3C273 for the *Astrophysical Journal Letters,* he suddenly recognized that the spectrum of the object could be explained only if it were that of hydrogen red-shifted by 16 percent. This red shift would imply that the object was at a great distance if it obeyed Hubble's Law. He immediately visited with Greenstein just down the hall. Greenstein suddenly recognized that the quasar spectrum he was working on, 3C48, had a red shift of 37 percent, implying that it too was at an immense distance. But if these objects were at these great distances, and are as bright as they appear, this implied that they are powerful sources of energy, a mystery still being investigated. (Photo by Darrel Hoff)

Discussion Questions

1. It was once quite uncertain whether quasar's red shifts were actually cosmological. What could be possible sources of the red shift if it was NOT due to Hubble's Law and the expansion of the universe?

2. If 3C273 were at the distance of the Andromeda Galaxy, what would its apparent magnitude be? Compare this to other celestial objects.

3. We seem to find quasars at large distances from us, but few, if any, nearby. Make as many arguments as you can for why this might be so.

APPENDICES

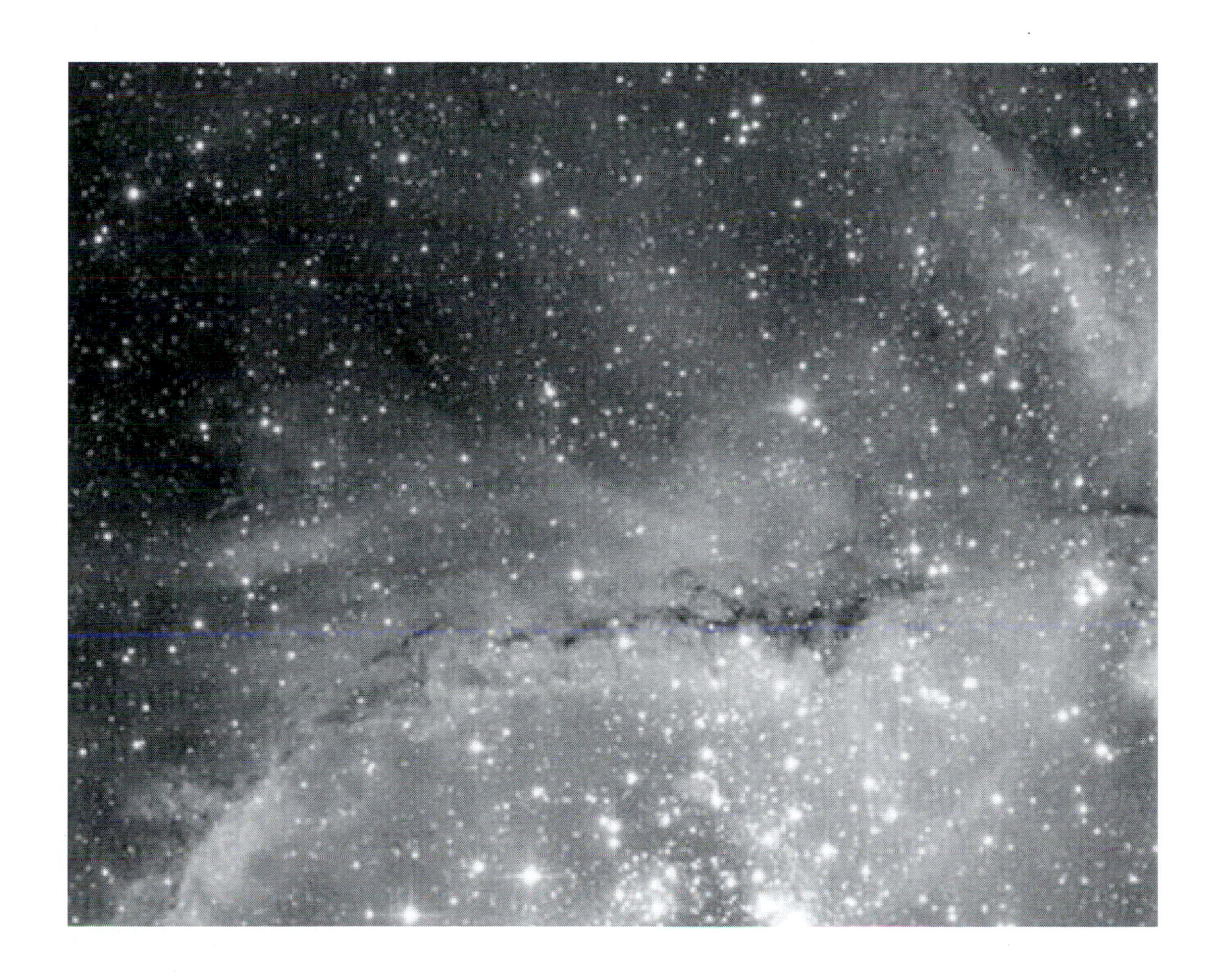

Appendix 1

How to Write Laboratory Reports

BASIC INFORMATION

At the beginning of each report give

1. Title of the experiment.
2. Date.
3. Laboratory partner(s), if any.

PURPOSE

Summarize in a few words the purpose of the exercise, including any modifications as the exercise progressed.

REFERENCES

If any references other than those given in the laboratory guide are used, list them.

PROCEDURE

Summarize in a few lines the basic procedure used. If any substantial changes from the procedure suggested in the laboratory guide were made, note them.

DATA

1. Label all data.
2. Give units for all data.
3. List data in tabular form.
4. If possible, take several readings for each data point and use the average value.

GRAPH

1. Plot each graph to as large a scale as is practical.
2. Title each graph.
3. Label the quantities plotted on each axis and label each curve if you have more than one.
4. Give units on each axis.
5. Where appropriate, draw a smooth curve or line through the data points. DO NOT CONNECT THE POINTS IN A DOT-TO-DOT MANNER. The actual data points do not have infinite accuracy, and thus may not lie exactly on the proper curve. Draw a smooth curve such that positive and negative deviations are about equal and such that the curve matches

the general trend of the data. This process averages the experimental fluctuations and the results deduced from the curve are usually more accurate than those deduced from individual measurements.

CALCULATIONS

1. Give calculations in a logical order down the page. Indicate the equation being used or the mathematical operation being done for each step.
2. Give units in each step of a calculation. Keeping close track of units may often help you to avoid errors.
3. If one method of calculation is repeated several times for different values, give a sample calculation and tabulate the results of the repeated calculations.
4. If a standard value is available for the quantity you have calculated, compare your experimental value to the standard and compute your percent of error. The percent of error is given by

$$\frac{\text{standard-value} - \text{experimental value}}{\text{standard value}} \times 100\%$$

Note that this error in your derived value is not the "experimental error" resulting from uncertainty in measurements. Percent uncertainty or discrepancy between two values is calculated in a similar manner.

CONCLUSIONS

Give a brief statement of your conclusions and final results.

DISCUSSION QUESTIONS

Answer all questions asked in the laboratory guide, as well as any others your instructor presents, as concisely and completely as possible. THINK before you write.

Appendix 2

How to Handle Data

The fundamental characteristic of science is the practice of observation and experimentation. In order to progress, science must be built on a foundation of measurements of physical quantities. These measurements result in data—often a string of numbers or pairs of associated numbers. The astronomer who has worked hard to design and carry out an observation must work equally hard at understanding what the numbers mean. This task involves squeezing the maximum amount of information from the data while taking care to understand the limits of the data and avoid introducing artifacts into that data. Graphing is a technique employed regularly by astronomers. Graphing helps one visualize the relationship between two variables and the exercises in this book often ask for a graph to be made. We will review a bit of graphing technique below and follow with a brief discussion of uncertainties in experiments.

GRAPHICAL ANALYSIS

In its simplest form a graph is a visual representation of how one measured quantity depends on another measured quantity. We plot one value, usually the "controlled" or independent variable, on the horizontal axis and the other value, the dependent variable, on the vertical axis. For instance, if we wanted to know how Wayne Gretzky's point production has depended on (ah!—this is the dependent variable) the year of his career, we could construct the graph shown in Figure A2-1.

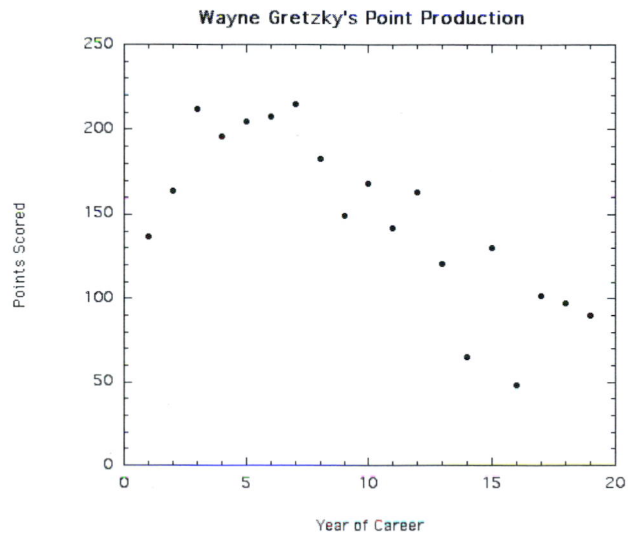

Figure A2-1. Wayne Gretzky's NHL points for each year of his career.

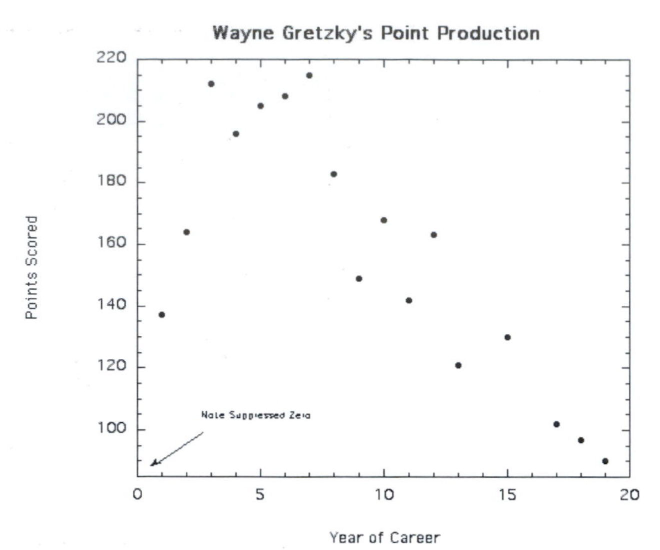

Figure A2-2. Gretzky's points with suppressed zero.

We see that just a few years into his career Wayne hit a general peak that lasted half a dozen years before his point production began trailing off. That is all we can say. We don't know whether NHL scoring in general has decreased over the time measured (it has) or how Wayne's production has depended on his changing cast of teammates, etc. The graph is good for showing general trends. Note that some points are anomolously low and that is clear when one views the entire graph. These low points arose when Wayne played fewer games, due to injury or labor unrest. This tells us that we might want to plot points per game instead of total points on the vertical axis. Our plot has led us to what might be a better way to view the data. We will not pursue this avenue, but if we did the new graph might prompt us to make yet another new graph. Data analysis is a learning process, in which the investigator must be open to new approaches.

It is often tempting to suppress a zero on one axis or the other. Since Gretzky always scored well above zero, we could truncate the vertical axis (ignoring the anomalous years) to make the graph shown in Figure A2-2. Suppressing a zero does highlight small features in the data but also exaggerates fluctuations. A casual observer of the second plot might be led to the conclusion that Gretzky's point production has varied much more dramatically over the years than it actually has. For that reason, it is best to produce full-scale graphs whenever possible. If you must suppress a zero, break the axis or make a note so that reader is alerted to this fact. Also, show both the suppressed and full-scale versions.

One of the things graphs can do is lead us to conclude whether or not two variables are linearly related or are related by any other function. In many people's minds, science is linked with mathematics. In part, this is due to the constant stream of numbers that make up the data. It has more to do, however, with the fact that mathematical laws can be formulated to "explain" existing results and predict the results of future measurements. When mankind first started measuring the universe around us, it would have been difficult to predict that we would find a multitude of mathematical laws overlaying the workings of nature. The fact that mathematical laws describe nature so well once led noted astronomer Sir James Jeans to proclaim, "God is a pure mathematician." One such law is $F = ma$, Newton's second law. This law expresses a linear relationship. For instance, if the net force on an object is doubled then the acceleration of the object is doubled. If we made a plot of force on the vertical axis and acceleration on the horizontal axis, we would get a straight line with a slope equal to the mass of the object. Slope is "the rise over the run," or the change on the vertical axis divided by the

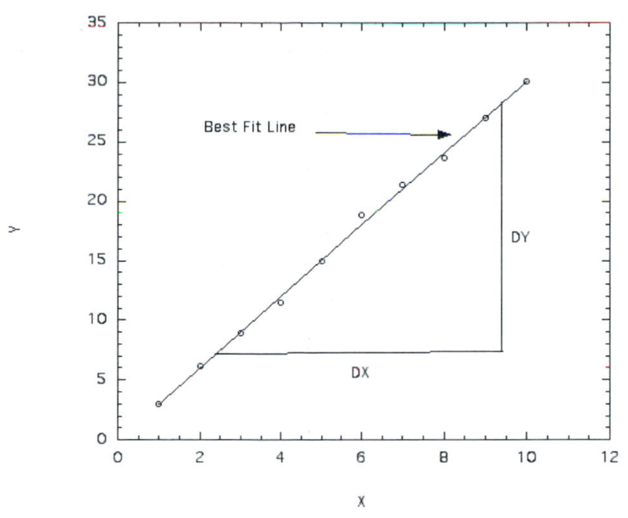

Figure A2-3. The slope of the best fit line is $\Delta Y/\Delta X$. Note that the X and Y scales are not the same.

change on the horizontal axis. The general equation of a line is $y = mx + b$, where m is the slope of the line and b is the intercept. An example of linearly related data is shown in Figure A2-3. When plotting data, y in the general line equation becomes whatever we plot on the vertical axis (F in the previous example) and x becomes whatever we plot on the horizontal axis (a in the previous example).

Many variables are linearly related (as F and a were), but many are not. The simplest of these other relationships are what astronomers call "power laws." Here we have $y = ax^n$. In this relationship, a is a constant and n is the power that relates the two variables. For a parabola, n is 2. Newton's Law of Universal Gravitation says that $F = cR^{-2}$, or that the force is proportional to one over the separation squared. Here n is -2.

Astronomers often use something called a log-log plot to determine whether or not two variables are related by a power law and what the power, n, is if they are. A mathematical property of logarithms, which we will not cover here, tells us that if one plots the log of the data values instead of the data values themselves, one will get a straight line if the values are related by a power law. The slope of that line will be the index, n. In practice, you can use special log-log paper, which takes the logarithms for you. Figure A2-4 shows a log-log plot of $y = x^3$. Most computer graphing packages will do linear and power law fits to sets of data. The program is not using the graph but the actual data to determine the best fit parameters.

Finally, a histogram is a number plot that is used frequently by astronomers. An example of a histogram could be the heights of all the people in your class. First measure everyone's height. On the horizontal axis of your plot divide the possible heights into "bins." Your bins could be 5 cm wide. Then count up the number of people with heights between 160 and 165 cm and plot that number in the bin from 160 to 165 cm. Do the same for 165 to 170 cm, 170 to 175 cm, and so on. An X-ray astronomer might measure the number of incident X-rays with energies between energy a and energy b, then the number between energy b and energy c, etc. The result is a spectrum as shown in Figure A2-5. One will typically vary the bin size so that the data aren't spread out into so many bins that the shape of the curve can't be seen and the data aren't all in just a few bins.

UNCERTAINTY ANALYSIS

We have presented the idea that astronomy, like all sciences, depends on measurements of physical quantities. One way to present these measurements is graphically, and analyzing the

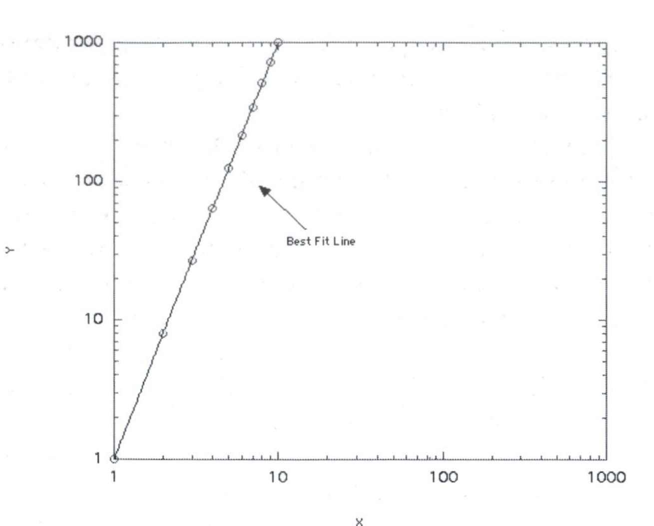

Figure A2-4. A log-log plot of $y = x^3$. The slope of the best fit line is the power, 3.

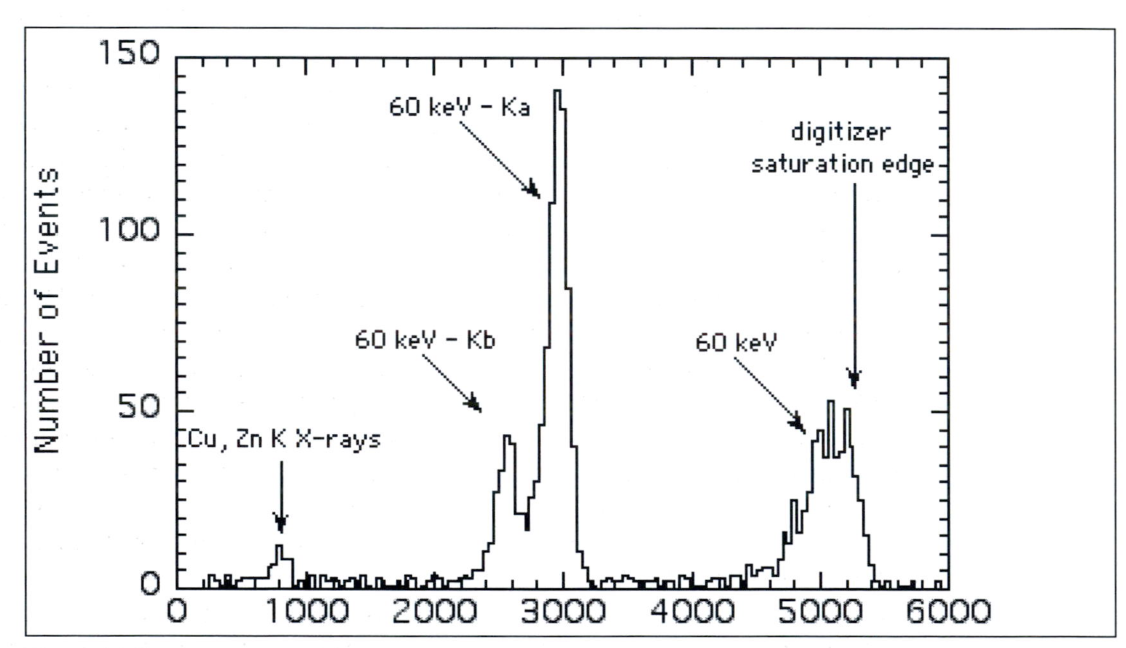

Figure A2-5. The energy spectrum of a radioactive source as measured by an X-ray detector. Each bin covers an energy range. We plot the number of events falling in that range, using a flat line from the beginning of the bin to the end of the bin.

data can yield an underlying mathematical relationship. As we proceed with such analysis, however, we must remember that all physical measurements have some uncertainty associated with them. All measuring devices possess an inherent precision limit and observing procedures may introduce uncertainties as well. For example, if you were measuring the height of your lab partner to make the histogram we discussed earlier, you might decide that you could only read the meter stick accurately to the nearest millimeter. It would be impossible to measure a height to better than 1 mm, but even reaching the 1 mm might be difficult. Care must be taken to ensure that the meter stick and subject remain parallel to one another. If the overall height of the person is greater than one meter (pretty likely) then you must mark a spot and slide the meter stick along. Can you mark the spot to within 1 mm? Maybe, but only if you work at it. It might turn out that you can measure the height with a precision of half a cm or even a full cm.

In any experiment or observation, the scientist must work to reduce uncertainties to a minimum in order to detect faint signals but must also work to understand what uncertainty remains. Any claim of a discovery is usually accompanied by a statement of the likelihood that the discovery is not an artifact. While it is not always easy to determine a precise uncertainty (indeed, this may be the most challenging aspect of some experiments) you should always think about what uncertainties are involved as you make measurements during various exercises. Try to decide what would cause the largest uncertainty and how great that uncertainty might be, just as we did with the height measurement. Most important is to remember that no quantity you measure is absolute.

Often there are many pieces that contribute to the uncertainty for an observation and the easiest way to measure that uncertainty is to make the measurement many times. In fact, the more times you measure a given quantity the better the precision of the measurement becomes. A perfect example of this can be found in the histogram shown in Figure A2-5. The spread of the peaks is caused entirely by uncertainty in the measured energy, uncertainty based solely on detector performance. The incident X-ray energies were identical inside each peak. Thus, the width of the peaks tells us the uncertainty in the measured energy of any one X-ray photon. Sometimes the energy we measure is a little higher than the actual energy; sometimes the energy we measure is a little lower than the actual energy. We get the curve you see. If we point the detector at a neutron star and measure one X-ray photon, we don't know if that measurement was a little high or a little low. We would have to report the energy of that X-ray photon as whatever we measured plus or minus the width of one of the peaks.

What saves us is the fact that we can measure the center of any one of those peaks pretty well and the center corresponds to the actual value. We have enough below and enough above the actual value that they average to the middle value with quite good precision. The more measurements we make the better we know what the average value is. One example of this type of measurement would be rolling a die. If you roll a die many times the average value will be 3.5. If you roll it just once, you certainly will not get 3.5. If you average two rolls, it's possible to get 3.5. In fact, 3.5 is the most likely value, but other values such as 4.5 are also fairly likely. As you roll the die more and more, the likelihood that you will measure an average, say, 20% or more away from the expected average grows smaller and smaller. You have some realistic chance of rolling two sixes in a row to yield an average of six but the odds of rolling 100 sixes in a row are so long against you that it is effectively impossible. Go ahead and try the experiment. Roll a die 10 times, then 100, then 1000.

In doing an astronomy observation we may be able to get many thousands of events or more in a given peak, so that we can make a relatively precise measurement. In order to do this we make telescopes or detectors bigger, observe longer, and lower the backgrounds in our detectors. We must be careful, though. If our instrument is changing slightly with time or the source is changing with time, those many thousand measurements we make might actually be measurements of different things and all our averaging just fools us.

So far we have been talking about statistical uncertainties. Another type of uncertainty is systematic. An example of a systematic error would be an error that comes from a meter stick that has shrunk so that each millimeter mark is really only separated by three quarters of a millimeter. We would systematically overestimate the size of every object we measured. This type of uncertainty can be difficult to eradicate. One possibility is to measure a known quantity. For example, if we had two objects of known length we could measure each of them with our faulty meter stick. We would find that our meter stick measured each of the objects to be 33% longer than their actual lengths. We could then scale all of our measurements with this meter stick by dividing by 1.33. Such corrections are rarely so simple and almost always lead to an increased measurement uncertainty.

A lot goes into making the measurements on which our understanding of the physical world is built. As you practice making measurements, keep these things in mind. Also, think about uncertainties and systematic errors the next time you hear a discovery reported. When you think about how difficult it can be to make these measurements and the truly vast distances over which astronomers attempt to make the measurements, you just might begin to appreciate something of what Einstein meant when he said, "The most incomprehensible thing about the universe is that it is comprehensible."

Appendix 3

Equipment Notes

Here are some informal notes on our experiences with equipment. The companies listed are not necessarily the best sources for any item, nor is this meant to be an exhaustive list. It is simply those items we are most familiar with. Prices given may not be up-to-date.

EXERCISE 1 VISUAL ASTRONOMY

1. Constellation charts. Extra SC1 and SC2 charts are available from Sky Publishing Company at $0.30 and less in quantity orders.
2. Pathfinder or star locators. Available from Edmund Scientific Co. at $2.75 (cheaper in quantity). A plastic planisphere (Phillips Planisphere) is available from Optica Company at about $9.95. (Be sure to state your latitude when ordering.) Lawrence Hall of Science distributes *Sky Challenger Star Wheels.* These are regular star finders; in addition insertable wheels are provided for locating binocular objects and for showing native American constellations. They sell for about $6.95. The *Observer's Handbook* is available for $15.00 from the Royal Astronomical Society of Canada. Guy Ottewell's *Astronomical Calendar* is available from Furman University or Sky Publishing for $15.00. Astronomical wall calendars are available from AstroMedia, Sky Publishing, or Hansen Planetarium for $8.95 to $9.95.
3. Binoculars
4. Small telescope (if possible)

EXERCISE 2 OBSERVING EXERCISES

1. Telescope
2. Additional moon maps. Moon maps and descriptions are found in many handbooks, field guides, and annuals.
3. *The Astronomical Almanac, Sky and Telescope, Astronomy,* or *Observer's Handbook* for planet positions.
4. Camera, film, and tripod (optional)
   You might want to suggest planets to observe or photograph during the semester or quarter. Venus is excellent (if visible in the evening!) and try to suggest the other planets as they retrograde (near opposition).
5. Plastic strip for measuring angles, star chart (optional)
6. Solar filter or cardboard ring and projection screen

Mylar material from camping "space blankets" can be used to make solar filters. It may require multiple layers. Carefully test the opacity before allowing student use.

7. Erfle eyepiece, if possible
8. The *Sky Challenger Star Wheel* (see Exercise 1 of this Appendix) provides an excellent reference set of the most easily located objects.

EXERCISE 3 STAR CHARTS AND CATALOGUES

1. Constellation charts. See Exercise 1 of this appendix.
2. The *Sky Gazer's Almanac* appears as an 11″ × 17″ color graphic in the January issue of *Sky and Telescope* magazine, as part of the *Current Guide to the Heavens* for $1.00 (less if ordered in quantity), and as a black and white wall poster for $4.95. All are available from Sky Publishing.
3. Rotating star and planet locators (see Exercise 1 of this appendix)
4. Various charts and catalogues listed in the exercise itself.

EXERCISE 4. ATMOSPHERIC EXTINCTION OF STARLIGHT

1. Graph paper or graphing software

EXERCISE 5 OBSERVING WITH SIMPLE TOOLS

Star counting. Cardboard mailing tube, 2- or 3-inch diameter, about 18 inches long works well. Star counting can also be done in the planetarium and toilet paper tubes work best here.

Simple angle measuring device.

1. Constellation charts (See Appendix 4.)
2. Plastic strip, about ¾ by 18 inches. Strips can be cut from nearly any flexible plastic materials, but clear strips are better. Ribbed plastic matting (often used in chemistry laboratories) and heavy duty plastic window coverings work well.

Building a pin-hole camera

1. Toilet paper tubes
2. Black construction paper
3. Rubber bands
4. Clear high wattage light bulb
5. 4″ × 5″ cut film (ISO 400 preferred)

Field of view of a telescope

1. Star charts or list of stars at various declinations
2. Small telescope with clock drive
3. Stop watch or digital watch with seconds displayed

EXERCISE 6. ANGULAR RESOLUTION

1. Shorter focal length lenses (5 to 10 cm works well)
2. Longer focal length lenses (20 to 40 cm works well)
3. Lens holders

EXERCISE 7 IMAGE SIZE—FOCAL LENGTH RELATIONSHIP

An optician's millimeter ruler and a vernier caliper are available from Edmund Scientific for $6.25 and $10.95, respectively.

EXERCISE 8 ASTRONOMICAL IMAGING

1. Films as listed in the experiment.
   - 100-foot rolls of film are available for loading of individual canisters.
   - Reusable canisters sell at most photo shops for about $1.00 each and daylight bulk film loaders can be purchased for about $25.00. Loading your own canisters provides substantial savings and large bulk rolls last for a considerable period of time if refrigerated. In addition, self-loaded 10-exposure rolls of film are often more convenient for astronomical work than longer commercial rolls.
2. Cameras. Any Polaroid with an electric eye works well. Older Bessler, Canon, Pentax, Nikon and Olympus SLR cameras work best for attaching to a telescope. Modern compact "user-friendly" cameras are not designed to have detachable lenses.
3. Barlow lenses. Available from most astronomical and optical supply houses. See *Sky and Telescope* or *Astronomy* magazines for advertisements.
4. Adaptors. Two basic adaptors are needed and are available from camera supply houses. You will need a "T-adaptor" to fit your particular camera. Prices range from $14.95 up to $32.50 depending on camera brand. You will also need a telescope adaptor to fit the diameter of your telescope eyepiece holder. These sell for $13.95 up to $19.95 depending on size of your telescope's eyepiece holder.

See our Web page for links to suppliers of digital imaging equipment.

EXERCISE 9 KIRCHHOFF'S LAWS AND SPECTROSCOPY

1. Spectroscopes. Holographic grating material for making your own is available in bulk ($5.00 for a 5″ × 9″ sheet and $25.00 for a 5″ × 6′ roll) and in 35-mm glass slide mounts ($30.00 for 10) from Learning Technologies. Cardboard spectroscopes with regular plastic gratings are available from various science supply houses for $30-$40 for class sets of 15.
2. Gas discharge tubes, holders, and power supplies. We suggest tubes of as many of the following gases as possible: H, He, Na, Hg, Ni, Air, O, H_2O, Xe, Kr. If possible, it is good to have spares, and you will want extras of several of them to use as "unknowns." Tubes are available from various science supply houses for $18-$26, with corresponding power supply for $140-$155. Fancy screw-in tubes are available from CENCO at $140-220 for the lamps and $500 for power supply and lamp case.
3. This setup can be used if gas discharge tubes are not available:
   Bunsen burners
   Platinum test wires
   Salts such as CaO, SrBr, KBr
   Dilute HCl (for cleaning test wires)

EXERCISE 10 ANGLES AND PARALLAX

1. Plastic strip, about ¾ inch by 18 inches. Strips can be cut from nearly any flexible plastic material, but clear strips are of some advantage. A ribbed plastic matting (often used in chemistry) has been used satisfactorily, and is very inexpensive.
2. Graph paper

EXERCISE 11 MEASURING DISTANCES TO OBJECTS OF KNOWN LUMINOSITY

1. Incandescent light bulb and socket
2. Digital voltmeter
3. Dimmer switch
4. Cadmium sulfide photocell or other light sensor
5. Meter stick

EXERCISE 12 DISTANCE TO THE PLEIADES
Ruler and compass (optional)

EXERCISE 13. DISTANCES OF CEPHEID VARIABLE STARS
1. Graph paper or graphing software

EXERCISE 14 GALACTIC DISTANCES AND HUBBLE'S LAW
Ruler (mm) or magnifier and reticle (see Exercise 19 of this appendix)

EXERCISE 15 DURATION OF THE SIDEREAL DAY
1. Camera, film, stopwatch, and tripod (optional)
2. Polar coordinate tracing graph paper (optional) or a compass

EXERCISE 16 LUNAR FEATURES AND MOUNTAIN HEIGHTS
1. Camera, film, and telescope to photograph moon (optional). It might be of value to plan a sequence of labs to photograph the moon in one session, develop and print the photographs in the next session, and to do this lab as a follow-up.
2. Commercial lunar photos (optional). Black-and-white wall poster-size photos of the first- and third-quarter moon are available from various science supply houses for $11.40-$16.25 per set.
3. Ruler and/or meter stick
4. *The American Ephemeris and Nautical Almanac* or *The Astronomical Almanac*

EXERCISE 17 ORBIT OF THE MOON
1. Camera, film, and tripod (if moon passes close enough to a bright star or planet to obtain your own photographs)
2. Tracing paper
3. Compass
4. Constellation chart (see Exercise 1 of this appendix)
5. Plastic strip, about ¾ inch by 18 inches. Strips can be cut from nearly any flexible plastic material, but clear strips are of some advantage. Ribbed plastic matting (often used in chemistry laboratories) and plastic window coverings have been used satisfactorily and are very inexpensive. See Exercise 2 for calibration techniques.

EXERCISE 18 DETERMINING THE MASS OF THE MOON
If a calculator-plotter or computer is available it can help relieve the tedium of plotting the many data points.

EXERCISE 19 EVIDENCE OF THE EARTH'S REVOLUTION
Millimeter ruler or magnifier with reticle. A flat plastic ruler works best. Various magnifiers and reticles are available from Edmund Scientific starting at $37.50.

EXERCISE 20 SOLAR ROTATION
1. Tracing paper
2. Ruler (mm) or magnifier and reticle (see Exercise 19 of this appendix)

EXERCISE 21 MEASURING THE DIAMETERS OF PLUTO AND CHARON
No special equipment needed.

EXERCISE 22 KUIPER BELT OBJECTS
No special equipment needed.

EXERCISE 23 DETERMINING THE VELOCITY OF A COMET
Millimeter ruler

EXERCISE 24 PROPER MOTION OF A STAR
1. This lab uses a piece of translucent graph paper found in Appendix 4.
2. Graph paper (mm)
3. Compass
4. Ruler

EXERCISE 25 SPECTRAL CLASSIFICATION
No special equipment needed.

EXERCISE 26 A COLOR-MAGNITUDE DIAGRAM OF THE PLEIADES
Camera (optional)

EXERCISE 27 SUPERNOVA 1987 A
Millimeter ruler or (preferred) vernier caliper (see Exercise 7 of this appendix)

EXERCISE 28 GALACTIC CLUSTERS AND HR DIAGRAMS
Graph paper

Note: This lab uses a transparency found in Appendix 4.

EXERCISE 29 THE DISTRIBUTION OF STAR CLUSTERS ON THE SKY
No special equipment needed.

EXERCISE 30 GALAXIES IN THE VIRGO CLUSTER
1. Palomar Sky Survey print ($+12°$, 12^h24^m). This is one print of a set of six available from the California Institute of Technology. A set of six different prints including the one referenced sells for $31.65 plus $3.50 for shipping. This set also includes other prints useful in the laboratory.
2. Ruler (mm) or magnifier with reticle (see Exercise 19 of this appendix)

EXERCISE 31 ABSOLUTE MAGNITUDE OF A QUASAR
1. Ruler (mm) or magnifier and reticle (see Exercise 19 of this appendix)
2. Graph paper

ADDRESSES OF SUPPLIERS REFERRED TO IN THIS APPENDIX
Astronomical Education Materials

AstroMedia Corporation
Box 1612
Waukesha, WI 53187–1612
(800) 533-6644
www.kalmbach.com

Astronomical Society of the Pacific
390 Ashton Ave.
San Francisco, CA 94112
(415) 337-1100

California Institute of Technology
Bookstore 1-51
Pasadena, CA 91125
(818) 356-6161

Discovery Corner
Lawrence Hall of Science
University of California Berkeley
Berkeley, CA 94720
(415) 642-1016

Government Printing Office
Washington, DC 20402
(202) 783-3238

Hansen Planetarium
1098 South 200 West
Salt Lake City, UT 84101–9917
(801) 538-2104

Science First/Starlab
95 Botsford Place
Buffalo, NY 14216

Sky Publishing Corporation
40 Bay State Rd.
Cambridge, MA 02138–1200
(617) 864-7360
www.skyandtelescope.com

Willmann-Bell
Box 35025
Richmond, VA 23235
(804) 320-7016 or (800) 825-7827
www.willbell.com

General Science Supplies

Central Scientific Co.
Now combined with
Sergeant-Welch (see below)

Edmund Scientific Co.
101 E. Gloucester Pike
Berrington, NJ 08007–3800
(800) 363-1992
www.edmundoptics.com

Fisher EMD
4500 Turnbury Dr.
Chicago, IL 60103
(800) 621-4769
www.fisherscientific.com

Frey Scientific
P.O. Box 8101
Mansfield, OH 44901–8101
(800) 225-3739
www.freyscientific.com

Sergeant-Welch
7350 N. Linder Ave.
Skokie, IL 60076
(800) 727-4368
www.sergeantwelch.com

Science Kit & Boreal Laboratories
7777 E. Park Dr.
Tonawanda, NY 14150–6784
(800) 828-7777
www.sciencekit.com

Vernier Software & Technology
13979 SW Millikan Way
Beaverton, OR 97005-2886
(888) 837-6437
http://www.vernier.com

Optical Companies

Astro-Physics
7470 Forrest Hills Rd.
Rockford, IL 61115
(815) 282-1513
www.astro-physics.com

Celestron International
2835 Columbia St.
Torrence, CA 90503
(310) 328-9560
www.celestron.com

Meade Instruments Corporation
6001 Oak Canyon
Irvine, CA 92618
(800) 626-3233
www.meade.com

Parks Optical
270 Easy St.
Simi Valley, CA 93065
(805) 522-6722

Questar
P.O. Box 59
Dept. 118
New Hope, PA 18938
(800) 247-9607
www.QuestarCorporation.com

Tele Vue Optics, Inc.
32 Elray Dr.
Chester, NY 10918
(845) 469-4551
www.televue.com

(For solar viewing only)
Corona
1674 Research Loop, Ste 436
Tucson, AZ 85710
(520-740-1561)
www.coronadofilters.com

Astronomical and Optical Supplies

Adorama
42 West 18th St.
New York, NY 10011
(800) 223-2500
www.adorama.com

Astronomics
680 SW 24th Ave.
Norman, OK 73069
(800) 422-7876
www.astronomics.com

Jim's Mobile, Inc.
810 Quail St.
Lakewood, CO 80215
(800) 247-0304
www.jimsmobile.com

Lumicon International
750 E. Easy Street
Simi Valley, CA 93065
(805) 520-0047
http://www.lumicon.com

Orion Telescope Center
2450 17th Ave.
P.O. Box 18155
Santa Cruz, CA 95061
(800) 447-1001
www.telescope.com

Miscellaneous

NASA Space Science Resource Directory
http://teachspace.org

Appendix 4

SC1-SC2 Charts, Star Finder Parts, and Translucent Graph Paper

SC002 CONSTELLATION CHART

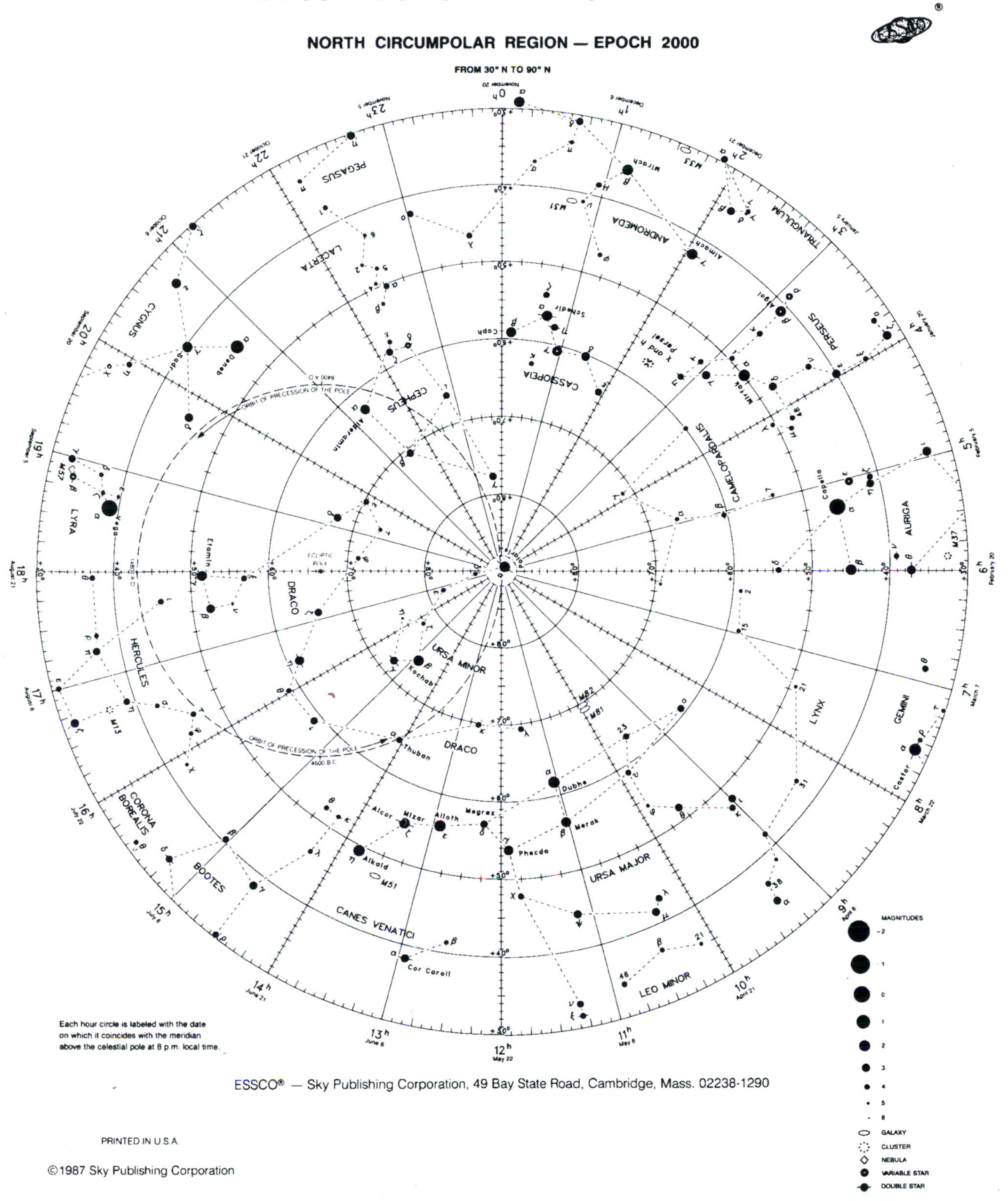

ESSCO® — Sky Publishing Corporation, 49 Bay State Road, Cambridge, Mass. 02238-1290

PRINTED IN U.S.A.

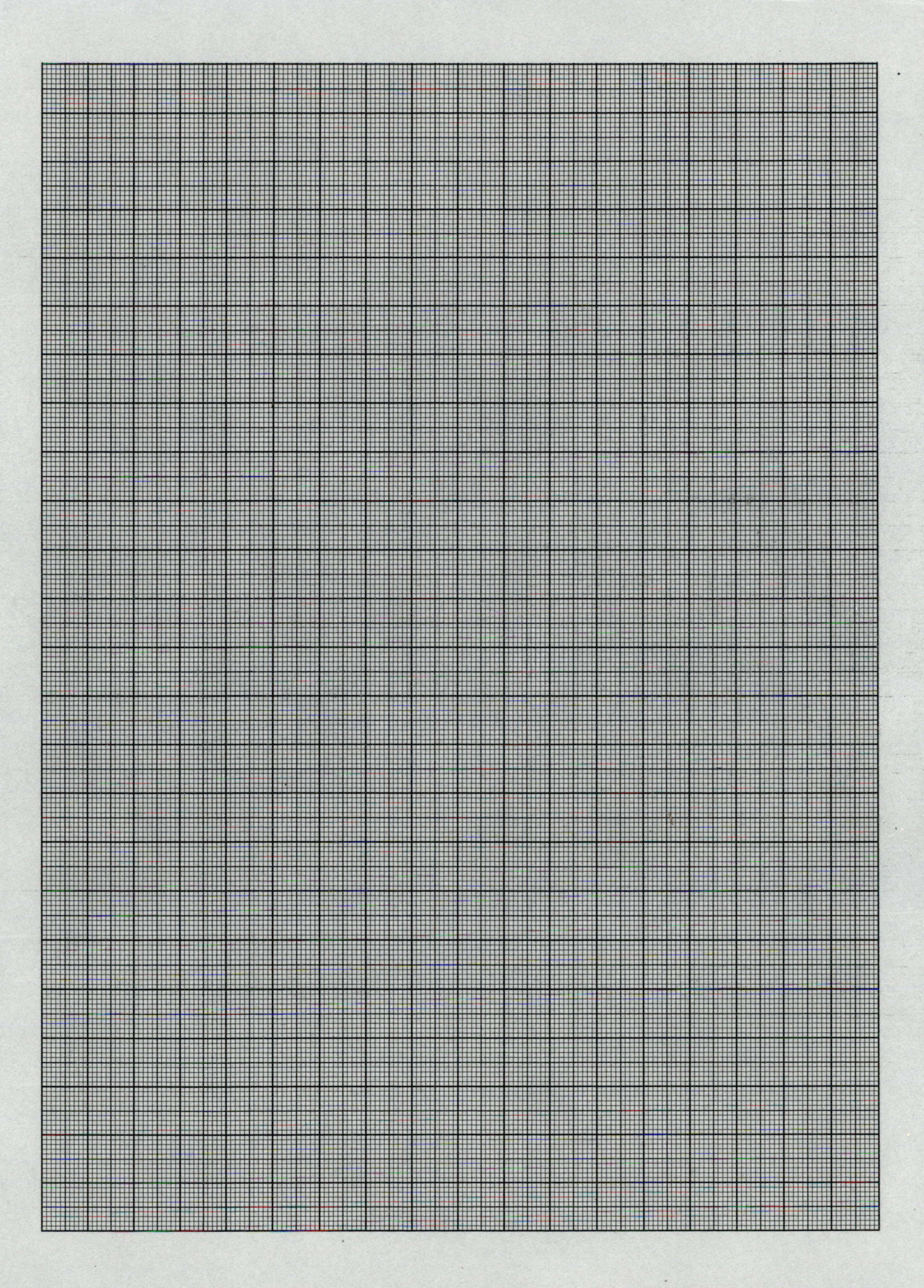

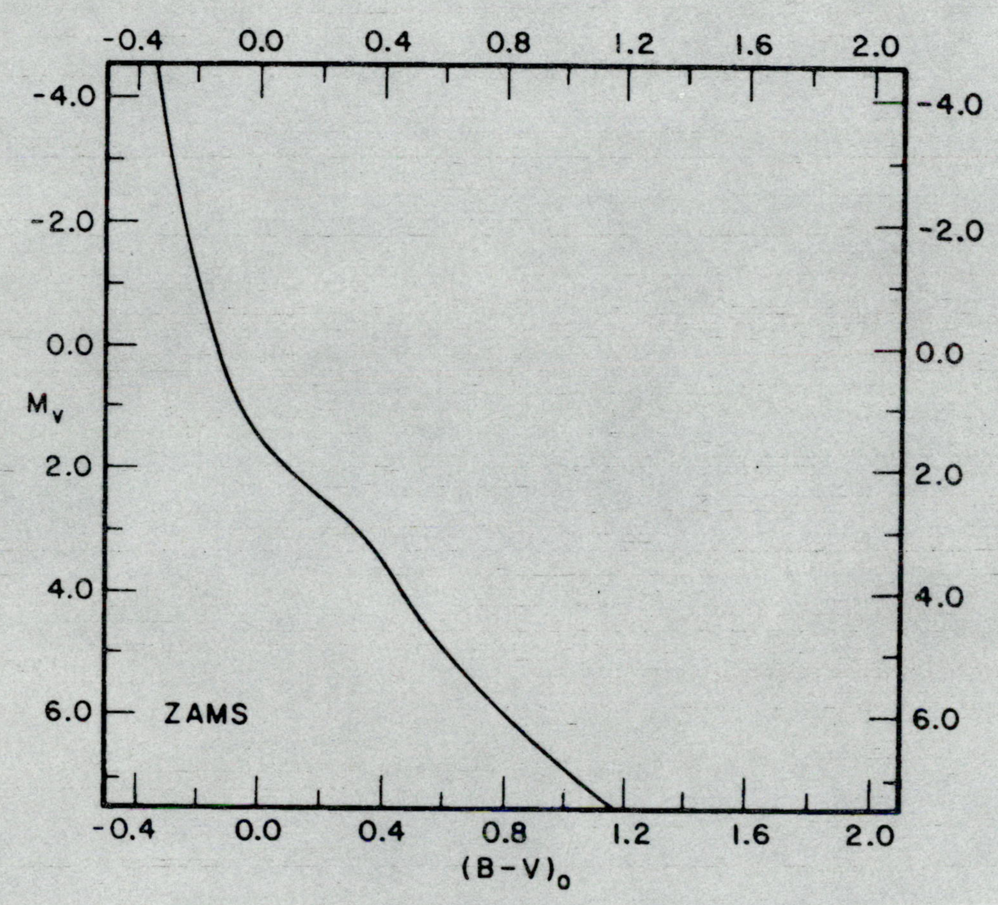

Figure 29-10. Zero age main sequence.

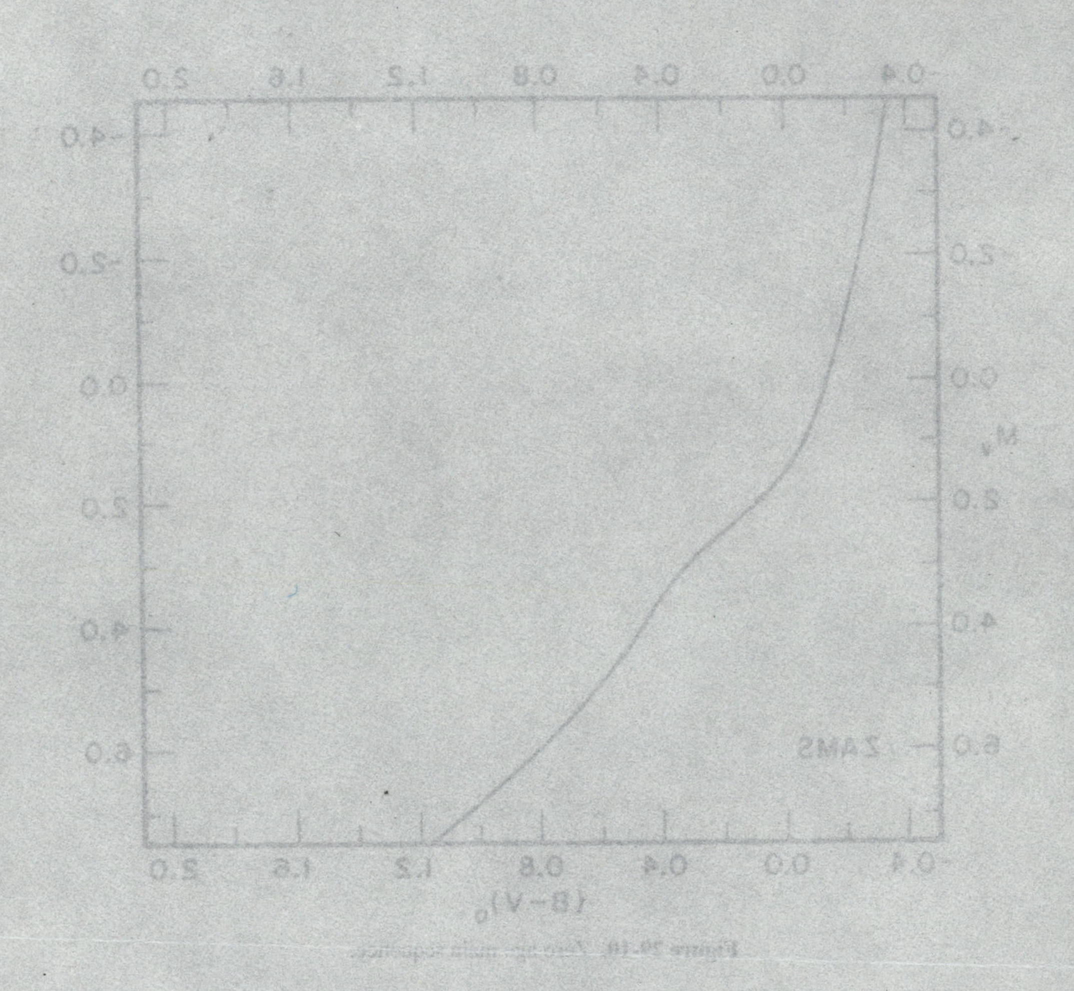

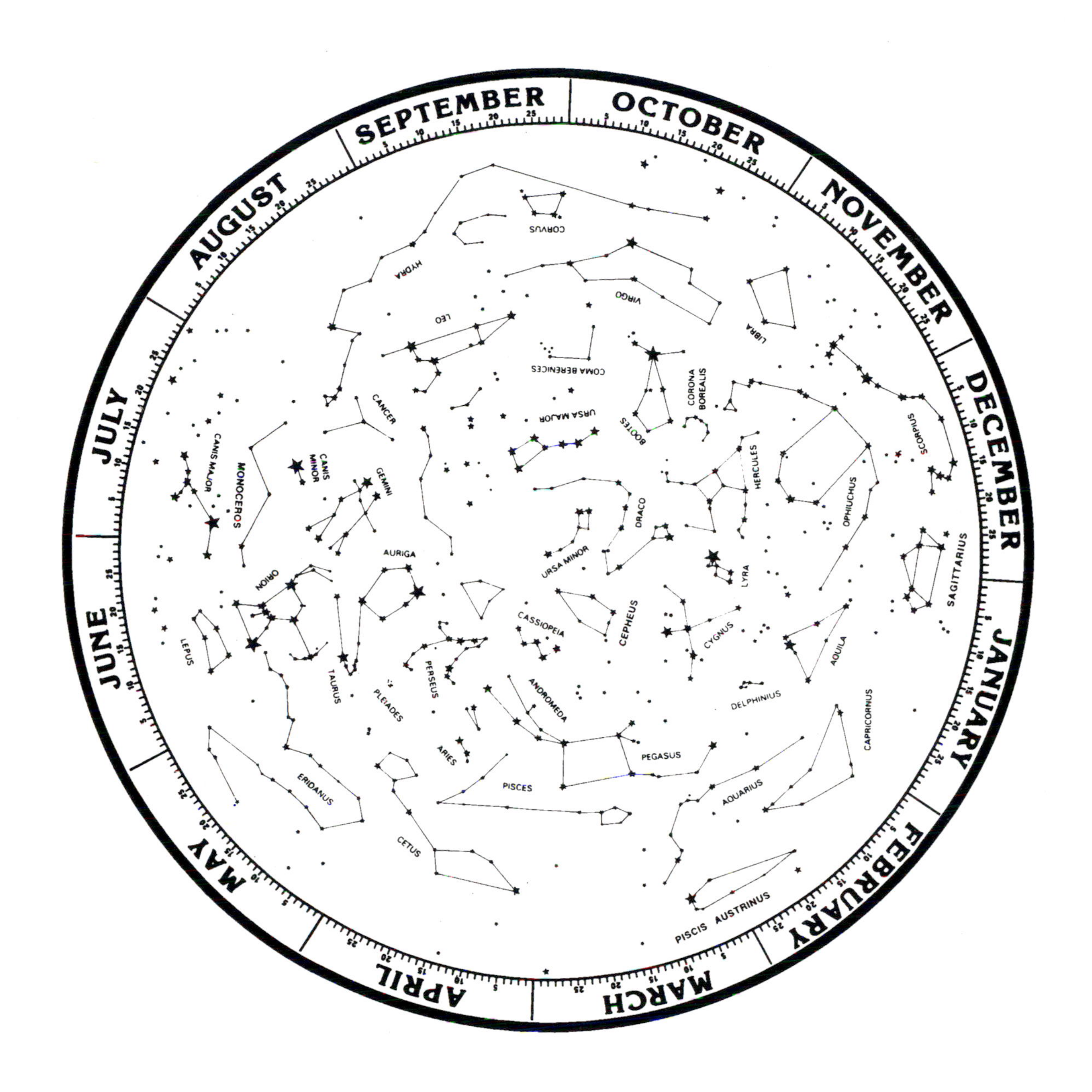
SEPTEMBER
OCTOBER
NOVEMBER
DECEMBER
JANUARY
FEBRUARY
MARCH
APRIL
MAY
JUNE
JULY
AUGUST
CORVUS
HYDRA
VIRGO
LIBRA
LEO
COMA BERENICES
CORONA BOREALIS
BOOTES
URSA MAJOR
CANCER
CANIS MAJOR
MONOCEROS
CANIS MINOR
GEMINI
SCORPIUS
HERCULES
OPHIUCHUS
DRACO
SAGITTARIUS
AURIGA
URSA MINOR
LYRA
ORION
LEPUS
CASSIOPEIA
CEPHEUS
CYGNUS
AQUILA
TAURUS
PLEIADES
PERSEUS
ANDROMEDA
DELPHINIUS
CAPRICORNUS
ARIES
PEGASUS
PISCES
ERIDANUS
AQUARIUS
CETUS
PISCIS AUSTRINUS

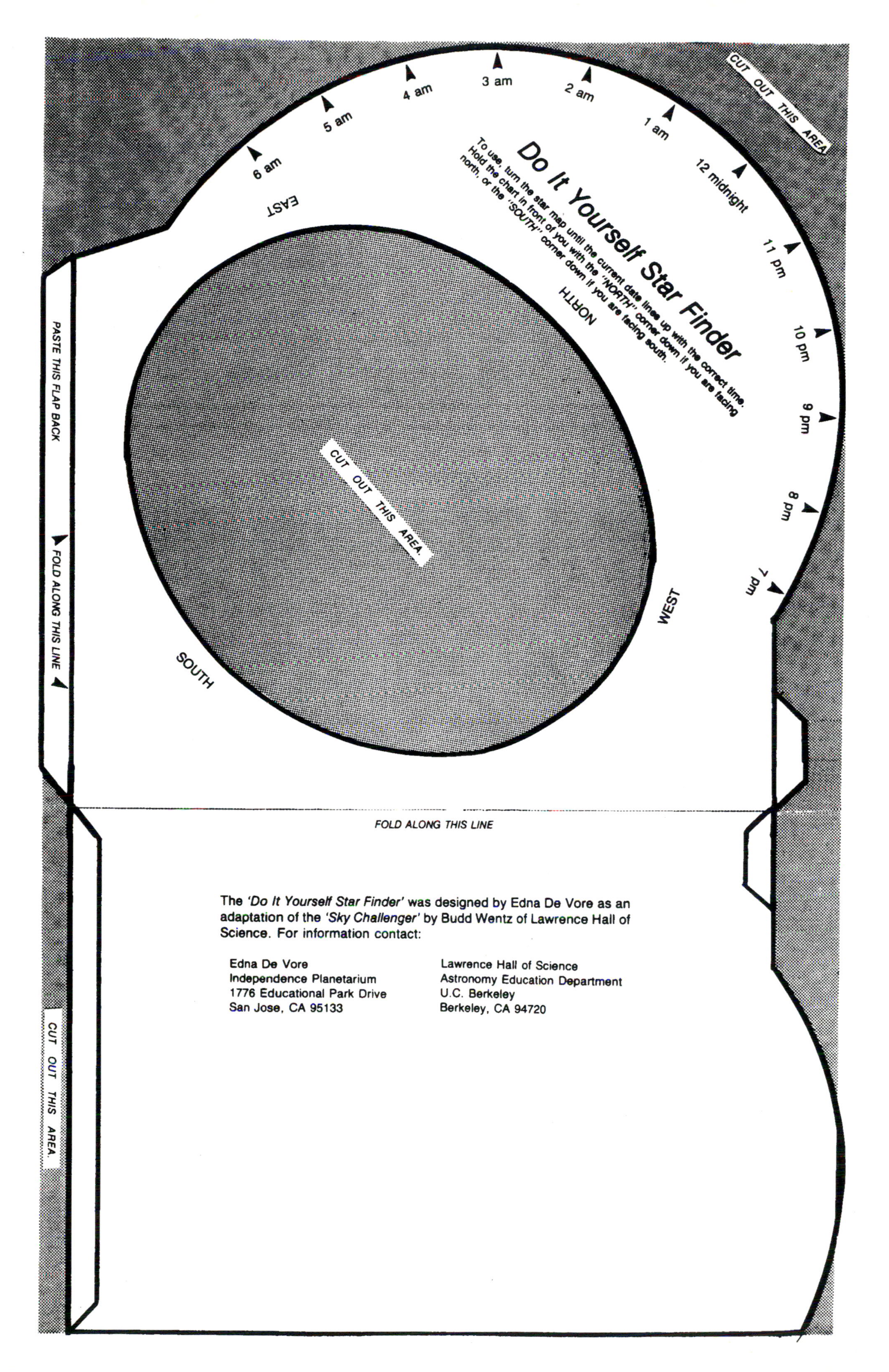

The *'Do It Yourself Star Finder'* was designed by Edna De Vore as an adaptation of the *'Sky Challenger'* by Budd Wentz of Lawrence Hall of Science. For information contact:

Edna De Vore
Independence Planetarium
1776 Educational Park Drive
San Jose, CA 95133

Lawrence Hall of Science
Astronomy Education Department
U.C. Berkeley
Berkeley, CA 94720

1. Experiment I: Visual Astronomy. Finding an observing site with dark skies in order to view the skies with the unaided eye can be rewarding particularly if one learns some of the brighter constellations. This 35mm photo is of the familiar winter constellation, Orion, the Hunter. Notice the subtle differences in the colors of the stars. In the upper left we find Betelguese, a cool, red star and in the lower right we find Rigel, a hotter, blue star. The middle 'star' in the sword is really a large glowing cloud of gas called the Orion Nebula (see photo 6, below). (Photograph by Darrel Hoff.)

2. Experiment 4: Atmospheric Extinction of Starlight. When the sun is either rising or setting, the thicker atmosphere provides extra filtering. At that time we also see the effect of the atmosphere's light scattering. The red rays pass through the atmosphere more readily while the blue wavelengths are scattered. This produces the effect of a reddened sun at those times. Also the atmosphere produces a refraction effect that results in a solar image that is demonstrably flattened. In this photograph the sun is setting in a cloud bank. Note the gradation in the sun's color and the flattened shape of the sun's image. (Photograph by Darrel Hoff.)

3. Experiment 2: Observing Exercises: Much greater details of solar system objects have been achieved as a result of the use of the Hubble Space Telescope and space probes. Contrast the detail of this photograph with what you might see with a small or modest aperture telescope. (1998 © Photo Disc, Inc.)

4. Experiment 2: Observing Exercises: A Voyager photograph of Jupiter shows much greater detail than can be achieved from the earth with even large-aperture telescopes. Being able to get above the earth's atmosphere means that we do not have to look through the ocean of air above us. (1998 © Photo Disc, Inc.)

5. Experiment 2: Observing Exercises. This 35mm photograph shows the face of Taurus, the Bull. The yellow star on the upper left of the "V" is Aldebaran. The tiny grouping of stars up and to the right of Taurus is the well-known open cluster of stars known as the Pleiades, or the Seven Sisters. (Photograph by Darrel Hoff.)

6. Experiment 2: Observing Exercises. A telescopic view of the Orion Nebula, found in the middle of the sword of Orion, shows that this region of the sky is not a single star but a complex of stars immersed in a cloud of glowing gas. (Courtesy of NASA.)

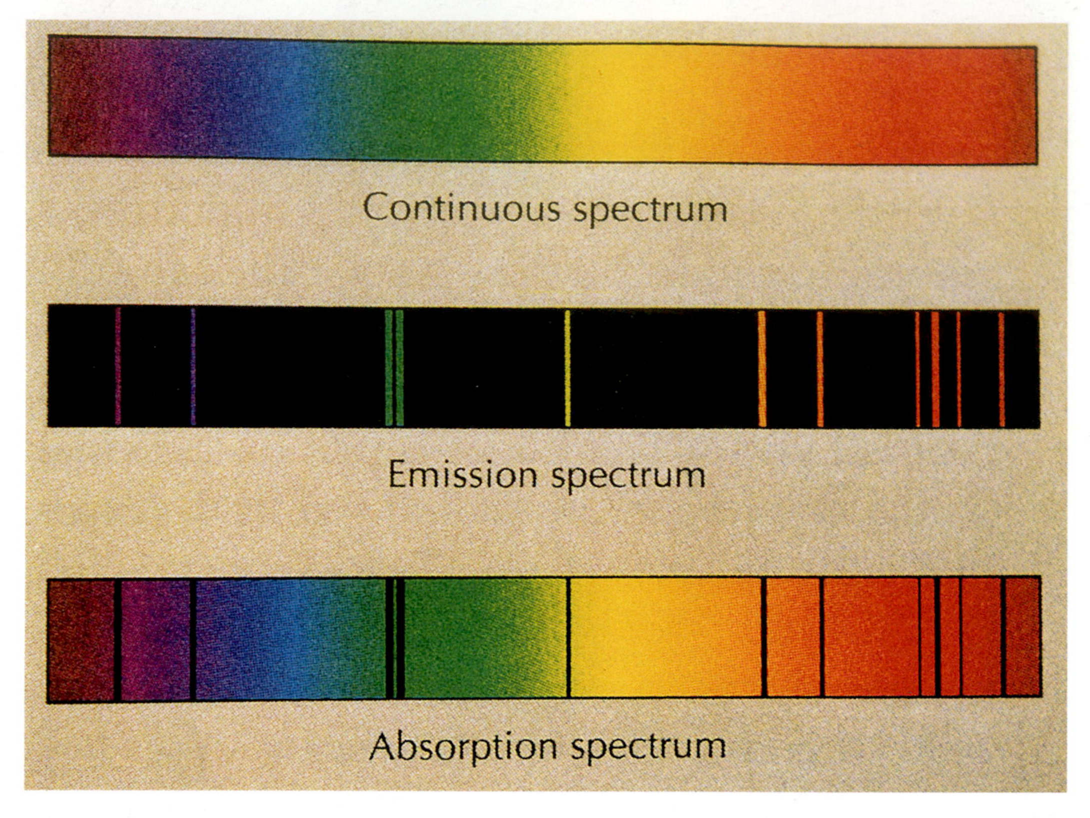

7. Experiment 9: Kirchoff's Laws and Spectroscopy. Three types of spectra.

8. Experiment 19: Evidence of the Earth's Revolution. Until the Space Age determination of the shape and motions of the earth was done using indirect measurements. The use of the Doppler effect to carefully measure the displacement of prominent spectral lines of a star located nearly along the ecliptic is one such example of the use of such indirect measurements. (1998 © Photo Disc, Inc.)

9. Experiment 29: The Distribution of Star Clusters on the Sky. Globular star clusters are groups of typically one hundred thousand to a million old stars. Approximately 150 globular star clusters are known to populate the halo of the Milky Way. Seen in the direction of the galactic center more than half them are found in three constellations: Ophiuchus, Sagittarius and Scorpius. (Courtesy of NASA/ESA. Hubble Space Telescope and the Gemini Observatory.)

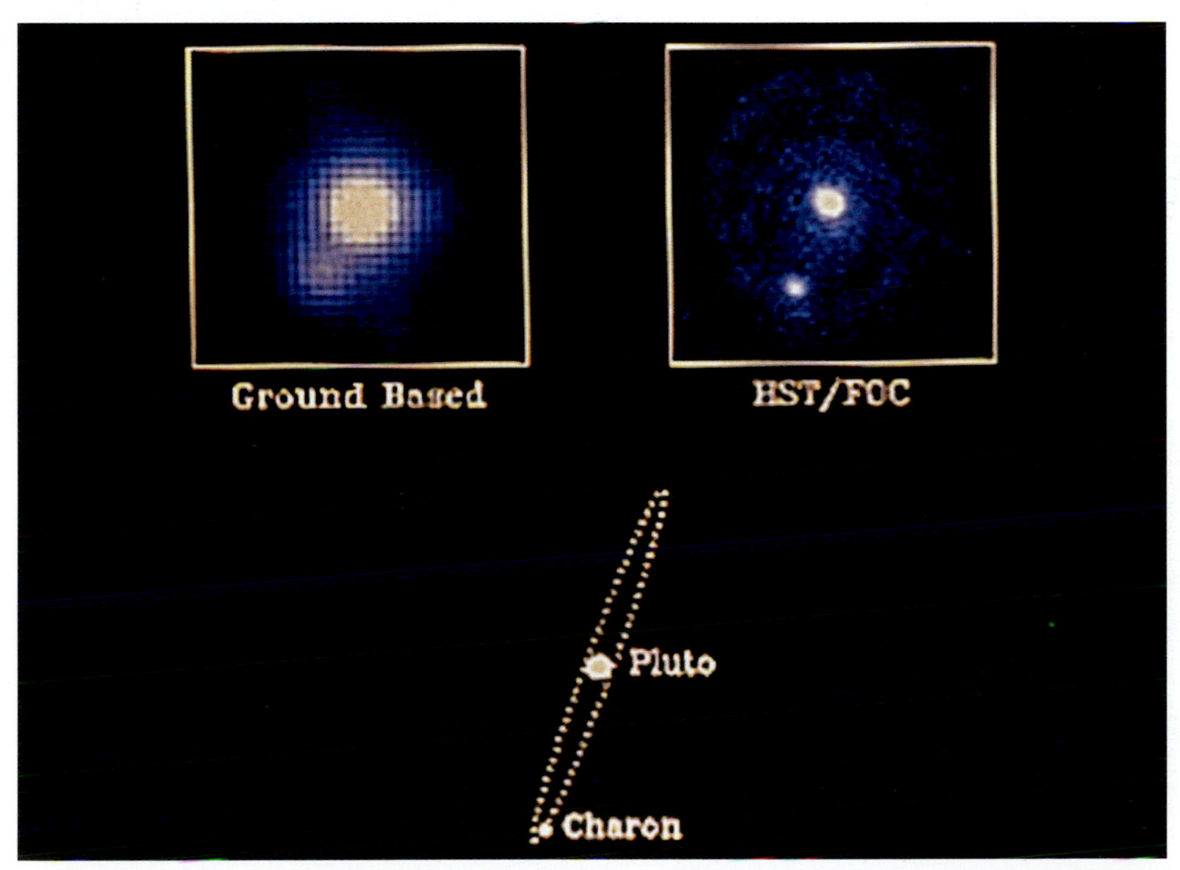

10. Experiment 21: Measuring the Diameters of Pluto and Charon. Christy discovered in 1978 that Pluto had a moon and that its orbit would put it into a series of mutual eclipses between 1985 and 1990. This discovery made possible the first accurate determination of the diameter of the planet—and its newly found moon. (NASA photograph using the Hubble Space Telescope.)

11. Experiment 23: Determining the Velocity of a Comet. The last decade and a half has seen a great deal of public interest in comets. The 1980s unspectacular return of Halley's comet was due to its orbital circumstances. On the other hand we were treated to the spectacle of Comet Shoemaker-Levy 9 impacting on Jupiter and the spectacular appearance of Comet Hale-Bopp in 1996 and 1997, which more than made up for any disappointment about Comet Halley. This picture was made with a 50mm SLR camera on April 17, 1997 when Comet Hale-Bopp was outbound in its orbit. (Photograph by Darrel Hoff.)

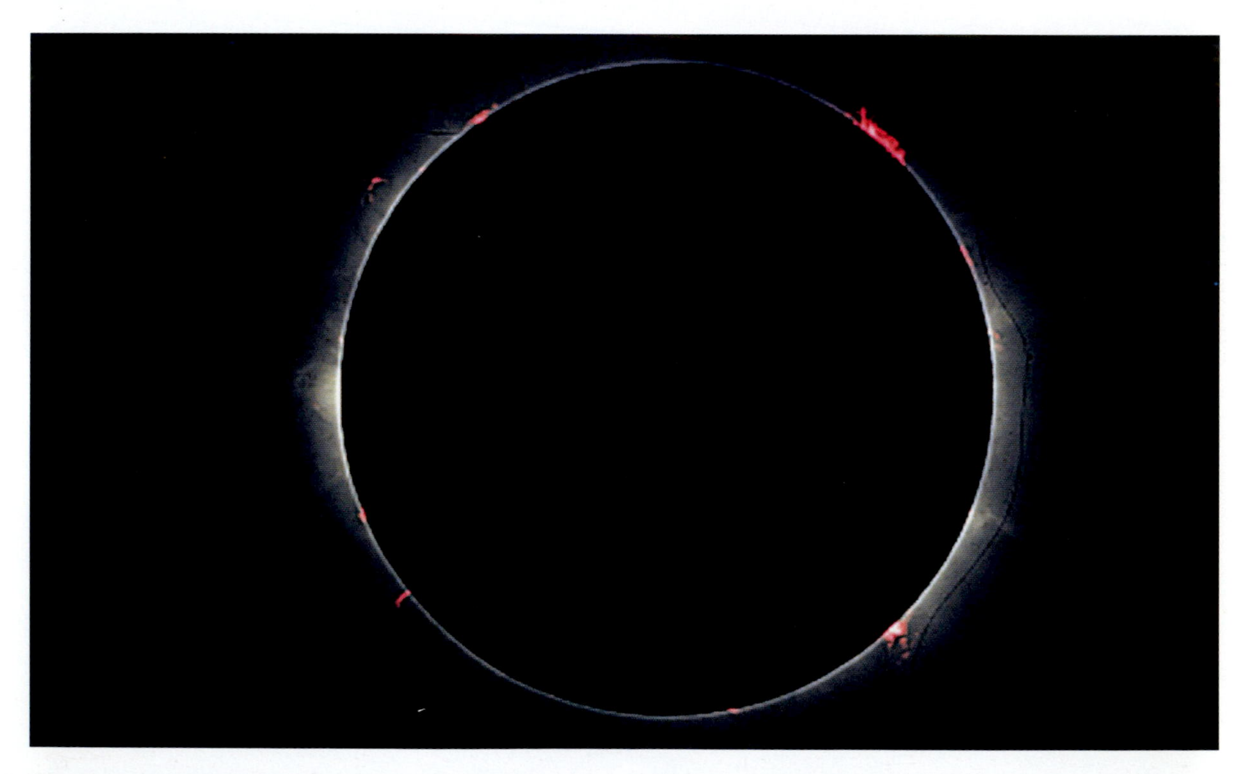

12. Experiment 20: Solar Rotation. The opportunity to directly observe a solar eclipse from any one location on the earth is rare. This image is of the total eclipse of February 26, 1976 as imaged with an 8-inch telescope from Bowbells, N.D. Fundamental measurements such as the sun's rotation period are determined from far more prosaic observations, such as using the apparent motion of sunspots or by measuring the Doppler shift of the edges of the sun's disk, as done in Experiment 24. (Photograph by Darrel Hoff.)

13. Experiment 25: Spectral Classification. The spectral classification scheme used today has a long and interesting history. In its modern form the Harvard spectral sequence (0, B, A, F, G, K, M) is really a temperature sequence with the hotter stars being 0 and B type stars and the cooler stars being K and M. Stars of intermediate temperature fall between these extremes. It is possible with long time exposure images to obtain a rough determination temperature using color film. This time exposure image taken near the celestial equator shows different color star trails indicating different color, hence different temperature, stars. (Photograph by Darrel Hoff).

14. Experiment 26: A Color-Magnitude Diagram of the Pleiades. This open cluster of stars is also known as the Seven Sisters, yet the average viewer can usually pick out only five or six stars. This cluster has served a vital role for astronomers as it is the closest such cluster in the sky. Astronomers use its group characteristics to search for relationships such as the relationship between spectral types and magnitudes. (Reprinted by permission of Palomar Observatory, California Institute of Technology.)

15. Experiment 27: Supernova 1987A. Supernova 1987A in the Large Magellanic Cloud (LMC) was interesting because it was the closest supernova in many centuries, and, because it occurred when it did, could be studied with a range of different types of modern instruments. Experiment 31 permits the student to establish what the apparent magnitude (brightness) the star had at its peak. Then, knowing the distance to the LMC, the student can calculate the star's absolute magnitude or its true brightness.

16. Experiment 14: Galactic Distances and Hubble's Law. The study of how far away galaxies are located became possible with the discovery of Hubble's Law in the 1920s. One of the closer galaxies is shown here in the photograph. Unlike the galaxies Hubble used to establish his law, the Andromeda Galaxy, M31, is one of the few galaxies approaching us. M31 is now believed to be 2.9 million light years from the earth. (Courtesy of NASA.)